Practical Agricultural Meteorology

NIPA® GENX ELECTRONIC RESOURCES & SOLUTIONS P. LTD.
New Delhi-110 034

Practical Agricultural Meteorology

Recommended/Prescribed as a text to students as per latest Dean Committees Report

A.K. SRIVASTAVA
Scientist
College of Agriculture, Tikamgarh-472001, Madhya Pradesh
&
P.K. TYAGI
College of Agriculture, Tikamgarh-472001, Madhya Pradesh

Jawaharlal Nehru Krishi Viswavidyalaya
Adhartal, Jabalpur, Madhya Pradesh, India

NIPA® GENX ELECTRONIC RESOURCES & SOLUTIONS P. LTD.
New Delhi-110 034

NIPA® GENX ELECTRONIC
RESOURCES & SOLUTIONS P. LTD.

101,103, Vikas Surya Plaza, CU Block
L.S.C. Market, Pitam Pura, New Delhi-110 034
Ph : +91 11 27341616, 27341717, 27341718
E-mail: newindiapublishingagency@gmail.com
Web: www.nipabooks.com

For customer assistance, please contact
Phone: + 91-11-27 34 17 17
Fax: + 91-11- 27 34 16 16
E-Mail: feedbacks@nipabooks.com

ISBN: 978-8119254-95-8

Composed and Designed by NIPA®.

Preface

Success or failure of a bio-system depends upon the weather conditions. The rate of growth and development of a bio-system is largely depending upon its ambient temperature and moisture. Agriculture is a field of the human activity which is especially sensitive to weather and climate. Each and every plant developmental phase is decided by meteorological parameters. Agricultural meteorology when properly applied can achieve the sustainability of agricultural production system through efficient management of agro-climatic resources and crop microclimate modification. The management and execution of agricultural farm practices, dairy farming, poultry, bee-keeping etc. are also weather dependent. The maximum efficiency of farm practices can be achieved with knowledge of prevailing and anticipated weather. Therefore, the knowledge of meteorological parameters and their influence on crop growth and yield is essential. Thus observations to be recorded extensively following standard methods and using well-calibrated instruments including the computation and the measurement techniques is absolutely essential.

In agrometeorological, teaching and research, meteorological data play very important role and applications. Though, practical manuals of agrometeorology are published by many organisations but they are lacking in materials, lucid presentation and good photographs of instrumets. Agriculture students are not finding their course materials in one book. Hence, a practical book on agrometeorology is very much needed, after fourth Dean Committee reports, for undergraduate agricultural students. The present book of Practical Agrometerology is prepared, which provides materials on all aspect of practical agrometeorology of data acquisition techniques, instruments and method of observation with precautions.

The book contains very useful and detailed information on practical aspects of agromet observatory and covers all major practical aspects of agromet observatory. This book is prepared with a series of pictures that illustrate the principles of meteorological parameters which are fundamental to the understanding of the subject by students and other readers. The book contains twenty-six chapters on installation of agro-meteorological observatory, measurement of rainfall and snowfall, air and soil temperature, humidity, wind, solar radiation, cloudiness, evaporation, evapotranspiration, sunshine duration, dew, pressure, photosynthetically active radiation (PAR), soil moisture content, leaf area index, transpiration, crop canopy temperature and computation of heat indices with solved examples. Chapters on agrometeorological data analysis, measurement of upper air observations, agro-meteorological data management and study of automatic-weather station have also been discussed in the book.

The book is written in simple language and easily understandable as the chapters have been elaborated with solved examples. We hope this book seems to be of great value not only to the students but also to the teachers/ scientists of various disciplines of agriculture like agronomy, entomology, pathology, nematology and other persons engaged in crop production. This publication will certainly make the students to compete for different examinations and to face the interviews with desired vigour and strength.

The encouragement and technical help rendered by colleagues and friends in bringing out the book 'Practical Agricultural Meteorology' is thankfully acknowledged. We are extremely thankful to the authors of various books, manuals and documents for getting useful material for inclusion in this publication.

We are gratefully acknowledge the India Meteorological Department, New Delhi, for utilising the synoptic maps, technical notes which are very helpful to write this book for the benefits of the students.

Finally, we acknowledge generously the Faculty of Agricultural Meteorologist of IARI, IMD, JNKVV and CCS HAU for their suggestions and encouragement to complete this book.

Tikamgarh (Madhya Pradesh) **A.K. Srivastava**
November, 2010 **P.K. Tyagi**

Acronyms Used in Manual

AET	Actual Evaporationspiration
Agromet	Agricultural Meteorology
APAR	Absorbed Photosynthetically Active Radiation
AWS	Automatic Weather Station
CC	Crop Coefficients
CCS HAU	Choudhary Charan Singh Haryana Agriculture University
CDR	Cyclone detector radar
DCP	Data Collection Platform
DCSTSS	Data Conversion, Storage and Transmission Sub-system
DWR	Doppler Weather Radar
ENSO	EI-Nino Southern Oscillation
ET	Evapotranspiration
ET_0	Reference Crop Evaporation
GDD	Growing Degree Day
GMT	Greenwich Mean Time
IARI	Indian Agricultural Research Institute
ICAR	Indian Council of Agricultural Research
IITM	Indian Institute of Tropical Meteorology
IMD	India Meteorological Department
IPAR	Incident Photosynthetically Active Radiation
IST	Indian Standard Time
JNKVV	Jawaharlal Nehru Krishi Vishwavidyalaya

LCD	Liquid Crystal Display
LMT	Local Mean Time
LWS	Leaf Wetness Sensor
MC	Meteorological Centre
MSL	Mean Sea Level
NCMRWF	National Centre for Medium Range Weather Forecasting
NWP	Numerical Weather Prediction
PAR	Photosynthetically Active Radiation
PET	Potential Evapotranspiration
PPFD	Photosynthetic Photon Flux Density
RH	Relative Humidity
RPAR	Reflected Photosynthetically Active Radiation
SMW	Standard Meteorological Week
Tc-Ta	Canopy Minus Air Temperature
TPAR	Transmitted Photosynthetically Active Radiation
USWB	United States Weather Bureau
UTC	Coordinated Universal Time
VP	Vapour Pressure
VPD	Vapour Pressure Deficit
WMO	World Meteorological Organization

Contents

❑❑❑

Chapter 1

Introduction

Agriculture encounters a complex dynamic system of natural conditions. The meteorological factors among these natural conditions play a pivotal role so far as agriculture is concerned. The rate of growth and development of a bio-system is largely depending upon its ambient temperature and moisture. The plant productivity per unit area, total production, quality of the produce, operational efficiency as well as econometrics are solely governed by the meteorological parameters. Therefore, their through understanding including the computation and the measurement techniques is absolutely essential. It is no less important to have sufficient workable information on weather phenomena as well as climatic resources to determine realizable potential and an appropriate solution for a number of problems faced in agriculture.

1.1. Agro-meteorological Data Collection

Crop growth is result of interaction between physical (weather or climate) and biological (genetic characteristics) elements. Hence, observations of physical and biological elements are essential.

Agro-meteorological data collection and its analysis are important for assessing the crop growth performances. Such data are also needed to assess the impacts of agricultural activities and processes on the environment and climate. Agro-meteorological data are used in agro-meteorological planning, characterization of agroclimate zones, crop-weather relationship, natural resources management, crop microclimate management and agromet advisory services. In agrometeorological studies physical as well as biological observations at local, regional and national level are important to know the association between plant and its environment.

1.1.1. Physical elements observations

Physical elements of climate and weather are observed to assist the management of agricultural activities. Such management includes determining the time, extent and manner of cultivation and other agricultural operations (sowing, harvesting, planting, application of biocides and herbicides, ploughing, harrowing, irrigation, suppression of evaporation, animal husbandry, etc.) and different methods of conservation, industrial use and transportation of agricultural products. Therefore, observation of important elements at standard time is a fundamental necessity.

1.1.2. Biological elements observations

In addition to the physical environment observation, the simultaneous biological observations both as individuals and as communities, is also required for agro- meteorological studies. The routine observations provided by climatological and agrometeorological stations should be accompanied by routine biological observations. For the best results, these observations should be comparable with those of the physical environment in extent, standard and accuracy. These observations are made to evaluate possible relations between the physical environment and biological growth, development and assessments of damage as well its performances.

1.1.3. Monitoring of atmosphere

Monitoring of atmosphere is a challenging task, need the different meteorological data at different time on horizontal as well as vertical scale, so that we can represent the atmosphere in three dimensional scales for it analysis. Monitoring of meteorological conditions is done at surface as well as upper air through a large network of observatories in India. Upper air data are of crucial importance for weather forecasting.

1.2. Variations in Weather and Climate

The variations of weather elements are brought about by the various climatic controls which operate in different combinations and with varying intensities. The variations of the weather or climatic elements are both temporal as well as spatial. The following are some of the most important climatic controls: latitude, longitude, altitude, distribution of land and water, high and low pressure systems, wind, air mass, ocean currents and mountain barriers.

1.2.1. Earth rotation and revolution

The earth rotates or spins about its axis in counter-clockwise at a speed of 465 metres per second at the equator. The rotation of earth causes the day and night at a place on the earth surface.

The rotating earth in space revolves round the sun, moving eastwards at mean speed of 29.8 kilometer per second. The time taken for one complete revolution of the earth round the sun is 365 days 5 hours, 48 minutes and 46 seconds, known as tropical year. The tilt of 23.5 degree of the earth's axis results in the unequal lengths of days and night at different positions of the earth in the orbit. The rotation, revolution and inclination of the earth produce different seasons.

1.2.2. Time

Meteorological observations all over the world are taken at fixed hours hence the time system employed must be on a universal basis. Otherwise the local time of observation would differ according to the place. Hence, it is essential to take into account the universal time system concept for recoding the meteorological observations.

1.2.2.1. Local time

The time at any point on the earth's surface, calculated by the position of the sun. This is also called solar time. In view of the rotation of the earth from west to eats on its axis, places lying on the same longitude will thus have the same noon and the same local time. As the earth rotate through 360° once in 24 hours, local time changes by an hour for every 15° longitude, or 4 minutes for every 1° longitude.

1.2.2.2. Standard time

The universal time system is based on the theoretical division of the globe, when we divide the globe by 24 circular lines meridianally, and thus producing 24 zones according to the hours in a day and each zone spacing 15 degree of longitude or approximately 7½ degree on each side of the standard meridian. These zones are known as Time Zones. The time established in each of the zone is called as "Standard Time". All places within the zones have the same standard time for particular region. Time zones are numbered 0 to 23, starting at the central zone of Greenwich and going east. The central meridian is the great circle connecting the north and south poles and passes through Greenwich in the United Kingdom (U.K.). The central meridian time of each zone is referred to as standard time. Large countries which stretch from east to west have more than one standard time. For example the old Soviet Union has 11 times zones and Canada has 6 time zones.

The central meridian adopted by India for the calculation of standard time is 82° 30′east longitude near Allahabad. Indian Standard Time (IST) is 05 hours 30 minutes ahead of GMT.

1.2.2.3. Greenwich Mean Time (GMT)

The time zone on central median, which passes through Greenwich (situated on zero degree longitude in U.K.) is known as Greenwich Mean Time (GMT). There is another time zone used in meteorology and known as universal time (UT). GMT is more or less the same as UT. The time used in the zones to those east of Greenwich is in advance of GMT and the time used in the zones to west of Greenwich is behind GMT.

1.2.2.4. Local Mean Time (LMT)

This is the local time based on the transit of the sun. LMT is the mean time determined with reference to the sun when sun rise at any place it is said to be at 7.00 0 clock. Locally, longitude 82° ½′ passing through Allahabad is called IST. LMT corresponding to IST varies form place to place depending upon the longitude of the place. LMT for a place can be calculated from the longitude of that place by the following relationship:

LMT = IST ± 4 (Difference in longitude of standard meridian and longitude of the place) minutes

If the station is to the west of reference longitude subtract 4 minutes for every degree from the IST, if the station is to the east of reference longitude, time has to be added to the IST.

Example : Calculate LMT for agrometeorological observations at Tikamgarh situated at a longitude of 79° 58' E.

Solution : The difference between longitudes of Allahabad and Tikamgarh is

= (82° 30' E- 78°58' E)

= 3° 32'

For each degree difference in longitude is 4 minute time difference. Therefore, time difference = 3° 32' x 4 = 14 minutes 28 seconds or 15 minute approx. Thus, 0700 and 1400 IST will be 15 minutes later at Tikamgarh or time of observation at Tikamgarh is 15 minutes a head of Allahabad.

1.2.3. Daylight (DL)

Daylight is defined as the interval between sunrise and sunset.

1.2.4. Geographical location of a place

Geographical location of a place on the Earth is indicated by latitude, longitude and altitude of that place as these are predominant factors that influence climate of a place. Any location on Earth is described by two numbers i.e., its **latitude** and its **longitude.** There are two **angles**, measured in degrees, "minutes of arc" and "seconds of arc." These are denoted by the symbols (°, ', "). For instance, 35° 43' 9" means an angle of 35 degrees, 43 minutes and 9 seconds. These are predominant factors that influence climate of a place.

1.2.4.1. Latitude

It is the angular distance (in degrees, minutes, and seconds) of a point north or south of the equator. These lines shown by horizontal lines on globe and are often referred to as parallels (Fig. 1).

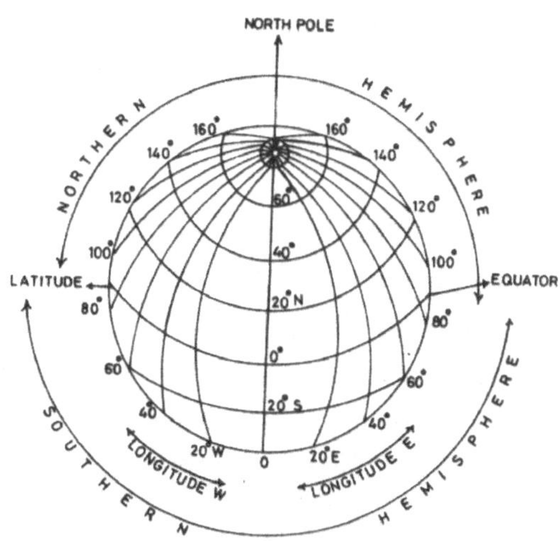

Fig.1 : Globe- demarcation features

1.2.4.2. Longitude

It is the angular distance (in degrees, minutes, and seconds) of a point east or west of the Prime (*Greenwich*) Meridian. These lines shown by vertical lines and are often referred to as meridians.

1.2.4.3. Altitude

The height of a place above a reference level, especially above mean sea level or above the earth's surface is called altitude .i.e. the vertical distance of a level, a point or an object from mean sea level.

1.2.4.4. Equator

It is an imaginary line lying mid way between the poles on the surface of the earth. It is also the longest circumference of the earth.

1.2.4.5. Poles

The two points at the northern and southern extremities of the earth is known as the North Pole and South Pole, which forms the ends of the earth's axis.

1.2.4.6. Hemisphere

One half of the earth's surface formed when a plane through its centre bisects the earth. The earth is usually divided into Northern hemisphere and Southern Hemisphere.

1.2.4.7. Above mean sea level

The term above mean sea level (AMSL) refers to the elevation (on the ground) or altitude (in the air) of any object, relative to the average sea level datum.

1.2.4.8. Mean sea level

The reference point used as a standard for determining terrestrial and atmospheric elevation or ocean depths is called the mean sea level (m.s.l.) and is calculated as the average of hourly tide levels measured by mechanical tide gauges over extended periods of time.

1.3. Components of Weather

As the atmosphere is in continuous motion, the state of the atmosphere varies with time and space. The condition of the atmosphere can be visualized through a set of parameters called the meteorological parameters or weather elements. A given atmospheric condition at a place at a particular time results from a complex interactions of different weather elements like rainfall, temperature, humidity, wind and radiation. Therefore, it is important to be aware of the interactions among these different elements of weather, when assessing the impact of one factor or when explaining the variation in it. For example, changes in air temperature will alter the air humidity and *vice-versa.* Thus, it is emphasized here that although each element is measured separately they are interrelated and ultimate weather condition is the sum total of all the parameters. Different weather parameters, particularly important for agriculture, that are regularly monitored are as follows:

1.3.1 Rainfall

Rainfall is precipitation (falling of any type of condensed moisture to the ground) in the form of liquid drops larger than 0.5 mm in diameter falling on the earth surface. Ordinary rain drop size varies from 0.5-4.0 mm in diameter. It is measured in term of depth of rain water (millimeters) on a solid surface provided that rain water is neither allowed to percolate nor runoff or loss in any form. It is measured by means of a raingauge. It consists of a funnel, a receiving jar with a narrow neck and handle, a cylindrical body, a splayed base which is fixed in the ground and a measuring cylinder. Rain water falling into the funnel is collected in the receiving jar housed in the external body. The water is transferred in to the measuring jar and the depth of rainfall is measured and expressed in units of millimetres. The raingauge should be installed on a level ground on a concrete foundation, so that the rim of the raingauge is exactly 30 cm above the ground. There are self recording rain gauges also which give a continuous record of rainfall during a day.

The intensity of rainfall can be worked out by the following formula:

$$\text{Intensity of rainfall (mm/hr)} = \frac{\text{Total rainfall}}{\text{Duration/ hr of rainfall}}$$

1.3.2. Temperature

Temperature is a relative term implying a degree of molecular activity or heat of a substance. It is a measure of intensity of heat and primary source of heat for earth and its atmosphere is the sun.

Atmospheric temperature is measured by means of thermometers in a special wooden box called Stevenson's Screen fixed at about 1.2 m above ground level, providing good ventilation besides sheltering the thermometers against direct sun's rays. Stevenson's Screen houses for four thermometers: maximum and minimum thermometers of horizontal type and two ordinary thermometers to serve as wet and dry bulb thermometers.

Maximum temperature is measured by maximum thermometer which is mercury in glass thermometer with a constriction in the capillary bore. Minimum temperature is measured by minimum thermometer which is an alcohol thermometer provided with an index in the stem. Dry bulb temperature is measured with a dry bulb thermometer and wet bulb temperature by wet bulb thermometer. Both are mercury in glass thermometers, but the bulb of the wet bulb thermometer is always kept moist by means of a moist wick.

Fahrenheit (F), Centigrade or Celsius (C) and Absolute or Kelvin (K) scales are used in air temperature measurement. Fahrenheit and Centigrade indicate the same temperature at -40 °C. The melting point of ice on the three scales is 32°, 0° and 273° and boiling point of water is 212°, 100° and 373°, respectively (Annexure-I).

1.3.3. Humidity

Atmospheric humidity or the water vapour content in the atmosphere is measured by psychrometer, a combination of dry and wet bulb thermometers. From the difference between the dry and wet bulb temperatures, using appropriate psychrometric tables, the relative humidity and the dew point temperature can be obtained.

1.3.4. Wind

Wind is air in horizontal motion in response to pressure gradient in the atmosphere. Wind measurement involves determination of direction of movement and speed with which it moves. Wind vane indicates the direction of wind. It is essentially a broad arrow head mounted on ball- bearing to enable the arrow to move freely in the horizontal plane about a vertical axis. The arrow head of the wind vane points to the direction from which the wind blows (windward direction). There are eight cardinal directions: N, NE, E, SE. S, SW, W, and NW.

Anemometers are used for measuring wind speed. Robinson's cup anemometer with three hemispherical metallic cups attached by means of metal arms mounted on vertical axis is commonly used in observatories. The cup wheel is pivoted at the centre and rotates as the wind blows. The revolutions are counted by means of a counter which directly gives the total distance wind traversed. Wind velocity is normally expressed in Kilometres per hour.

1.3.5. Sunshine

Bright sunshine hours are measured by means of an instrument known as Campbell-Stokes sunshine recorder. It consists of a glass sphere mounted concentrically on a spherical bowl. The sun's rays are focused by the glass sphere on to a card held in the groves cut of the bowl. Three sets of cards are used and each card is marked with hourly intervals. Depending on the total length and time of burning on the card, the number of bright sunshine hours in any day can be obtained.

1.3.6. Clouds

Clouds are condensed (or sublimation) form that usually results from lifting processes. A cloud forms when an ascending column of moist air is cooled below its dew point and the moisture condenses on minute particles called nuclei, which are present in the atmosphere. Estimation of cloud amount consists in estimating how much of the total apparent area of the sky to the nearest eighth, is covered with cloud. It is done by observing the sky with naked eye and is expressed in "oktas", ranging from 0 to 8. The amount of cloud is to be taken as "zero" only if the sky is completely cloudless and "eight" when the sky is completely overcast without gaps of any kind.

1.3.7. Evaporation

Evaporation refers to the amount of water transferred to the atmosphere in the form of water vapour. The rate of evaporation is defined as the amount of water lost by evaporation from unit area of a surface in unit time and is expressed in depth unit mm per day just like rainfall.

The class A pan evaporimeter consists of a large cylindrical pan made of copper having 122 cm diameter and 25cm depth is used for evaporation measurement. The pan is painted tin inside and white outside. The pan rests on a white painted wooden grill kept on the level ground in the observatory. A still well made of copper and having a vertical rod fixed at the centre of its base, is provided inside the pan. The tip of the rod which is tapered acts as reference point. The pan is filled with water till the water level just touches the tip of the rod. The pan is also provided with a standard wire mesh covering the evaporimeter. For measuring evaporation a graduated measuring cylinder is also supplied with the instrument. It has a scale of 0-20 cm engraved inside. Each time when an observation is taken, measured quantities of water with the help of measuring cylinder are added to the pan such that the water level is made up to the fixed point. The amount of water added is equal to the amount of evaporation taken place since the last observation. In case of rain when the water level in the evaporimeter rises above the pointer, measured quantity of water is removed from the pan using the measuring cylinder, till the water level again comes to the reference level. Knowing the rainfall occurred and the water removed from the pan, evaporation can be obtained.

1.3.8. Dew

Dew is the condensation of water vapour from adjacent air layers upon surfaces, cooled by radiation loss. Dew forms on the ground and on exposed objects, during winter, especially on clear and calm nights. Dew is measured by means an instrument called Duv Devani dew gauge. It consists of a set of four rectangular wooden blocks painted red. The four wooden blocks are kept at four heights on the stand 5, 25, 50 and 100 cm above the ground. The dew gauge is exposed in the open on a stand in the evening. Next morning, at sunrise, the dew formed on the dew gauge is observed and compared with a set of standard photographs contained in the dew album, which represent different dew scales from 0 to 9. Thus knowing the dew scale number, the water equivalent in mm, can be determined.

1.4. Types of Observatories and Stations

Meteorological stations of different types are installed by India Meteorological Department (IMD) in different parts of the country and a list of them is given below (Table 1).

1. Surface Observatories
2. Upper Air Observatories
3. Aeronautical Meteorological Observatories
4. Climatological / Hydrological Stations

5. Special Stations;
    a. Storm detecting radar stations;
    b. Cyclone Warning radar stations;
    c. Radiation observatories;
    d. Ozone observatories;
    e. Environmental Meteorological stations;
    f. Satellite Meteorological Stations;

1.4.1. Surface observatories

Surface Observatories are of mainly of three types, Meteorological, Agro meteorological and Hydro meteorological observatories. At meteorological observatories only daily weather data from a place is collected by IMD personnel. At agro meteorological observatories data of weather as well as soil and crops are collected. They are maintained by the State Agricultural and Irrigation Department, Agricultural Research Institutes and research farms and Agricultural university or Zonal and Regional Research centers. These observatories supply meteorological data to the India Meteorological Department.

The hydro meteorological observatories recording mainly precipitation data, were set up in connection with flood forecasting, river basin studies compilation of rainfall statistics for use in construction of dams, railways and road bridges, evaluation of snow-melt in Himalayan rivers and water balance and glaciological studies.

1.4.1.1. Evaporation observatories

Standard USA open pan-evaporimeters are installed at 238 observatories to measure evaporation.

1.4.1.2. Evapotranspiration Stations

Evapotranspiration in plants is measured by means of lysimeters at 42 stations to determine the water requirement of important crops. The network of 42 stations is the representative of different agro-climatic regions in the country.

1.4.1.3. Soil Moisture Stations

Fifty-five soil moisture stations are at present functioning. Most of these stations are maintained by IMD personnel.

1.4.1.4. Dew Fall Measuring Stations

Besides rainfall, dew fall is one of the secondary source of moisture available to the crops. It plays a significant role in plant growth, particularly in arid and semi-arid regions. It is measured from September to April by 76 observatories.

1.4.2. Upper air observatories (Radiosonde, Radio wind and Pilot Balloon Observatories)

The upper air meteorological data collected from all over the country are used on real time basis for operational forecasting. There are at present 65 Pilot Balloon Observatories and 39 Radiosonde Observatory in India.

Table 1 : Types of Observatories and their total number

S.No.	Types of Observatories	Number
1.	Surface Observatories	559
2.	Agrometeorological Observatories	264
3.	Hydro meteorological Observatories	701
4.	Evaporation Stations	238
5.	Evapotranspiration Stations	42
6.	Soil Moisture Recording Stations	55
7.	Dew-fall Recording Stations	76
8.	Ships of the Indian Voluntary Observing Fleet	203
9.	Pilot Balloon Observatories	65
10.	Aviation Current Weather Observatories	73
11.	Radiosonde Observatories	39
12.	Radiation Observatories	45
13.	Aviation Forecasting Offices at National & International Airports	19
14.	Storm Detection Radar Stations	17
15.	Cyclone Detection Radar Stations	10
16.	Snowgauge Stations	21
17.	Data Collection Platforms through INSAT	101

1.5. Inspection and Supervision of Observatories

Agricultural meteorological stations maintained by the National Meteorological Service should be inspected at least once a year to check instruments exposure and to ensure that observations conform to the appropriate standards and that the instruments are functioning correctly and are calibrated at the required times. If agricultural meteorological observatories are made by other authorities, cooperatives or special agreements should be made to ensure adequate supervision and maintenance of the instruments, including calibrations and exposure.

1.6. Meteorology and Agro-meteorology

Meteorology means the physics of the lower atmosphere and studies the individual phenomena of the atmosphere. Climate is an active factor in the physical environment of all living things. Climatology is the science that seeks to describe and explain the nature of climate, why it differs from place to place, and how it is related to other elements of the natural environment and to human activities. Climatology is concerned with the discussion of the component element of climate as well as the factors which determines and controls its distribution.

Meteorology and climatology are closely related. Meteorology is mainly concerned with physical analysis of the individual parameters, while climatology aim a systematic study of the climate and its distribution on the earth's surface.

Agrometeorology is the study of those aspects of meteorology which have direct relevance to agriculture. The task of an agro meteorologist is to apply every relevant meteorological skill to help the farmers to make the most efficient use of his physical environment for improving agricultural production both in quality and quantity. It is an interdisciplinary science and studies the interaction of physical, chemical and biological sciences.

1.7. Scale of observations

For the study of three dimensional atmosphere, we have to consider the scale in time and space (horizontal and vertical) both. In agricultural meteorology observations are required on the macro-, meso- and micro-scale. Local observations of physical parameters from fixed points as well regional and synoptic range of area are required to represent the weather at a particular time. For bio-meteorological research the micro-scale observations are often required.

1.7.1. Meteorological scales

In the study of the atmosphere, meteorology can be divided into distinct areas of emphasis depending on the temporal scope and spatial scope of interest. In the timescales of hours to days, meteorology separates into micro, meso, and synoptic scale meteorology. The classification of motion scale used in meteorology is given in Table 2.

Table 2 : Scales of motion used in meteorology

System	Approximate dimensions		
	Horizontal Scale (km)	**Vertical Scale (km)**	**Time (hours)**
Macro-scale			
a) Planetary waves	5000	10	200 to 400
b) Synoptic perturbation	500 to 2000	10	100
Meso-scale	1 to 100	1 to 10	1 to 10
Micro-scale	< 1/10	<1/100	1/10 to 1/100

1.7.2. Spatial scales

Meteorology can be divided into distinct spatial scale of interest into micro, meso, and synoptic and global scale meteorology.

Micro-scale meteorology is the study of atmospheric phenomena of about 1 km or less. Individual thunderstorms, clouds, and local turbulence fall within this category. Meso-scale meteorology is the study of atmospheric phenomena that has horizontal scales ranging from micro-scale limits to synoptic scale limits and a vertical scale that starts at the Earth's surface to the lower section of the stratosphere. Meso-scale timescales last from less than a day to weeks. The events are thunderstorms, squall lines, fronts, extratropical cyclones, and sea and land breezes. Synoptic scale meteorology is generally large area in horizontal coordinates and with respect to time. It includes events like extratropical cyclones, troughs and ridges, and frontal zones. All of these are plotted at a given time on weather maps.

Global scale meteorology is study of weather patterns related to the transport of heat from the tropics to the poles. Also, very large scale oscillations are of importance. Those oscillations have time periods typically longer than a full annual seasonal cycle, such as ENSO etc.

1.8. Space-time Scale in Agrometeorology

Like weather systems in meteorology there is no such systems found in agrometeorology, but have crops which have space (crop field) and time (crop season). Weather parameters averaged over space and time has their impact on the growth and development of crop. One day is too small and one month is too large duration for any crop. Therefore, Food and Agricultural organisation (FAO) recommended two time scales weekly and 10 days values. Therefore, the weekly or decadal values of weather parameters either totals or averaged over week or decal is used in agrometeorology.

1.9. Meteorological Organisation

As atmospheric conditions maintain a continuum across the globe and therefore, the weather monitoring has to be at the International level. This calls for a uniform procedures to be adopted by all the national agencies, assigned with the weather recording and archiving. A free flow of data between different countries is also required to put all the information, collected by individual nations, together to get a complete picture of atmospheric conditions in different parts of the globe. There is an International agency, World Meteorological Organization (WMO) situated at Geneva, Switzerland to coordinate these activities at the International level and WMO provides the framework for such international cooperation. At the national level, India Meteorological Department (IMD) is the sole agency to look after the weather data acquisition and other weather and hydrological related services to the nation.

1.9.1 World Meteorological Organization (WMO)

The World Meteorological Organization is an intergovernmental organization with a membership of 187 Member States and Territories. Established in 1950, WMO became the specialized agency of the United Nations for meteorology (weather and climate), operational hydrology and related geophysical sciences in 1951. Since its establishment, WMO has played a unique and powerful role in contributing to the welfare of humanity.

The WMO was entrusted with responsibilities to facilitate world-wide cooperation in the establishment of networks of stations for the making of meteorological as well as hydrological and other geophysical observations related to meteorology, and to promote the establishment and maintenance of centres.

1.9.2. India Meteorological Department (IMD)

India Meteorological Department (IMD) is the sole national agency for establishment of network of meteorological/agrometeorological observatories in the country. IMD is also responsible for meteorological data collection, its archival, analysis and weather forecasting in India. It was established in 1875 and Mr. H. F. Blanford was the first imperial meteorological reporter to the Government of India. Headquarter of IMD is New Delhi and main office of Meteorological Data storage and Agrometeorological work is in Pune. IMD functions under the Ministry of Earth Sciences, Government of India.

For the convenience of administrative and technical control, there are 6 Regional Meteorological Centres, each under a Deputy Director General with headquarters at Mumbai, Chennai, New Delhi, Kolkata, Nagpur and Guwahati. Under the administrative control of Deputy Director General, there are different types of operational units such as Meteorological Centres at state capitals, Forecasting Offices, Agrometeorological Advisory Service Centres, Flood Meteorological Offices, Area Cyclone Warning Centres and Cyclone Warning Centres. The country is divided into 35 meteorological subdivisions.

IMD has its 28 Meteorological Centres (MC) at the State capitals. IMD has strong international linkages with World Meteorological Organization (WMO). IMD has been issued national, state and district level medium range forecast and agromet advisory bulletins twice a week on every Tuesday and Friday with help of its 130 Agro-meteorological Field Units which are established in different Agricultural universities and ICAR Institutes.

1.9.2.1. Agrometeorological division

It is a part of IMD started functioning in 1932 at Pune under Dr. L.A. Ramdas. In India, there are about 264 agromet observatories under SAU's, Research Institutions and private agencies. At about 50 agromet observatories observations on growth of the crops like paddy, wheat, jowar, cotton, sugarcane and at 100 observatories qualitative observation on pests and diseases are carried out. It helps to train observer, conduct research in agro-meteorological discipline. It imparts technical guidance, in the establishment of observatory.

1.9.2.2. Network of observatories (IMD)

The observations are generally established throughout the country by IMD on surface, water and space as may be required with an aim to understand climate, climatic changes and to forecast weather for the safety and prosperity of mankind. IMD maintains a countrywide network of observational stations viz. 559 surface meteorological observatories, 701 hydrometeorological observatories, 65 pilot balloon stations, 39 radiosonde/rawinsonde observatories, 264 agrometeorological observatories, 45 radiation observatories from where weather reports are received at Pune through the Regional Telecommunication Hub based at New Delhi.

1.9.3. National weather modeling organizations

1.9.3.1. India Meteorological Department (IMD)

It is the national meteorological service of the country and the principal government agency in all matters relating to meteorology, seismology and allied subjects. IMD has operational Numerical Weather Prediction (NWP) system for forecasting weather in the medium range (3-10 days in advance) scale. It issues the 5 days district level forecast for the whole country every days for the benefit of the human society. It has its forecasting Offices at New Delhi and Pune.

1.9.3.2. National Center for Medium Range Weather Forecasting (NCMRWF)

The National Center for Medium Range Weather Forecasting (NCMRWF) was established by the Government of India in December 1988. It headquarter is at NOIDA (UP). The NCMRWF functions under the Ministry of Earth Sciences, Government of India. The mandated of NCMRWF is to develop operational Numerical Weather Prediction (NWP) system for forecasting weather in the medium range (3-10 days in advance) scale. It has the first super computer of CRAY series, which was used solely for medium range weather forecasting since 1989 in India.

The National Center for Medium Range Weather Forecasting (NCMRWF) is premier institution provide medium range weather forecasting through deterministic methods. A number of global, regional and meso-scale numerical forecast models are being run in real time by this Centre. These are global modes T-80/L18 - horizontal resolution of 150 km, T-170/L28 with 75 km horizontal resolution, Embedded Regional Spectral Model with 50 km grid resolution and Meso-scale Models MM5 with 30 and 10 km resolution and ETA Model with 48 km resolution.

1.9.3.3. Indian Institute of Tropical Meteorology (IITM)

The Indian Institute of Tropical Meteorology (IITM) was established in 1962 as Institute of Tropical Meteorology (ITM) at Pune as a distinct unit of India Meteorological Department for the study of the fundamental atmospheric problems in the tropical region. The Institute was transferred into an autonomous organization on 1st April 1971 under the name of Indian Institute of Tropical Meteorology. The IITM functions under the Ministry of Earth Sciences, Government of India. The IITM has developed techniques and expertise in specialized area of research such as studies on weather forecasting, climatology, monsoon studies, climate modeling, hydrometeorology, weather modification, atmospheric chemistry, atmospheric electricity, cloud physics and instrumentation for observational studies.

1.9.3.4. CSIR Centre for Mathematical Modelling and Computer Simulation (CMMACS)

The Council of Scientific and Industrial Research (CSIR) established in 1988 the "CSIR Center for Mathematical Modelling and Computer Simulation", briefly called *C-MMACS*. The Centre is located in National Aerospace Laboratories (NAL), Bangalore. A number of activities have now developed into important areas of research and development on modelling of hazard environments, pollution and climate change. This centre also started work on evolutionary features of cyclones and developing mathematical models for early warming of cyclones. Recently this center working on ocean modelling, and started the experimental forecasts of all India summer Monsoon rainfall by using mathematical simulation models.

Questions for Study

Q.1. What are the differences between meteorological and agrometeorological scale?

Q.2. Distinguish between Greenwich Mean Time (GMT) and Indian Standard Time (IST).

Q.3. Write about the weather and its components.

Q.4. What do you understand by meteorological organisation?

❑❑❑

Chapter 2

Installation of Agro-meteorological Observatory

Since meteorology and climatology are primarily observational sciences, adequate care has to be taken for getting most representative and accurate observations of weather parameters for their efficient application. This depends on site of observatory, maintenance of observatory, accuracy of instruments, timely measurements of weather parameters, knowledge and responsibility of the observer.

Monitoring of weather conditions, in a country like India, which is climatologically so diverse because of its unique geographical situation is a challenging task. It is done through a large network of observatories of different types. India Meteorological Department (IMD), Pune, is entrusted with the collection and keeping the record of weather data collected from different stations and also to train the man power in this specialized area. Weather stations installed in different parts of the country are of different types.

2.1. Observatory

An observatory is a specially designed station or place where observing instruments are installed. In the observatory, regular and simultaneous records of the weather data are made by physical measurements using various standard instruments and methods at specific hours recommended by the authorized National (IMD) and International agencies (WMO).

Meteorological observatory is a place where several weather-recording instruments are installed to record data on weather elements. Whereas, crop weather observatories or Agro-meteorological observatories are specialized observatories to record data on crop environment required for understanding the crop-climate interactions. Network observatories/ stations established in India are as under:

- Agro-meteorological observatories
- Evaporation Stations
- Evapotranspiration Observatories
- Soil Moisture Observatories
- Dew Fall Recording Stations
- Pilot Balloon Observatories
- Marine Meteorological Observations
- Flood Meteorological Officers for Flood Forecasting Observatories
- State Raingauge Stations

2.1.1. Agro-meteorological observatory

Agro-meteorological observatory is the station wherein physical elements of climate and agricultural/biological elements, or both are observed using standard instruments under standard exposure at specified time. These observatories are equipped with both eye reading and self-reading instruments, which measures various phenomenons's to describe the state of atmosphere near and around the surface of the earth. This is the zone in which most biological activities takes place. These instruments provide the measurement of weather elements, which characterize the weather at particular instant of time, near to the ground surface.

Agro-meteorological observatories are installed either by Agricultural Meteorological Division (IMD) Pune or by SAU (State Agriculture University), National Research Institutes or by any other organization.

2.1.2. Classification of agro-meteorological observatory

On the basis of number of instruments, type of instruments (Essential and Optional) and observations, World Meteorological

Organization (WMO) has divided agro-meteorological observatory into following four categories *viz.*, Principal or Class-A Observatory, Ordinary or Class-B Observatory, Auxiliary or Class-C Observatory and Specific purpose or Class-D agro-meteorological observatories. Each class of observatory requires essential and optional instruments as given in Table 3.

Table 3 : Types of agromet observatory, their essential and optional instruments

Type	Observation	Essential instruments	Recording type	Optional instruments
Principal Agromet observatory Or Class A	All meteorological observations	Thermometers (Max., Min., DB, WB), Rain gauge, Anemometer, wind vane, Evaporimeter, Sunshine Recorder, Dew gauge, Thermograph, Radiometers (Pyranometer and Pyrheliometer)	Self recording, direct recording and eye reading	Lysimeter, Portable potentiometer, Thermopiles,
Ordinary Agromet observatory or Class B	Almost all meteorological observations	Thermometers (Max, Min, DB, WB) and Soil thermometers Rain gauge, Anemometer, Wind vane, Evaporimeter	Self recording, direct recording and eye reading	Sunshine recorder, Dew gauge, self recording rain-gauge, Thermograph, Hygrograph
Auxiliary agromet observatory or Class C	Selected meteorological observations	Thermometers (Max, Min, DB, WB), Rain gauge	Direct recording and eye reading	Wind vane, Anemometer, Dew gauge
Specific purpose Agromet observatory	Only specific meteorological observation	This type of observatory is set up for the observation of one or som e specified weather parameter on temporary basis	Eye reading, direct recording	Rainguage

2.1.2.1. Principle or class-A observatory

The weather elements are observed at every three hour or twice a day. These observatories are equipped with eye reading, self-recording and direct recording instruments.

2.1.2.2. Ordinary or class-B observatory

The weather elements are observed twice daily and telegraphed to the different forecasting centers. These observatories are not completely equipped as Class-A type and the observations are recorded with eye reading and self as well as direct recording instruments.

2.1.2.3. Auxiliary or class-C observatory

The observations are taken only once in a day and telegraphed to different forecasting centers. The observations are recorded with eye and direct reading instruments.

2.1.2.4. Specific purpose or class-D observatories

These types of observatory are set up for the observation of one or some specified weather parameter on temporary basis. For instances, some observatories record rainfall and temperature and once a week data is telegraphed to forecasting centers. Some observatories record only rainfall and they telegraph messages only on the amount of rainfall if any during past 24 hrs.

2.2 Instruments

2.2.1. Requirements of instruments

Accuracy, reliability, robust construction simplicity of design, convenient in operation and maintenance and durability are the important requirements of the instruments.

2.2.2. Form of instruments

The instruments may be divided into two basic forms:

1. Direct-reading or non recording and
2. Self recording or automatic instruments

2.3. Observations

The observations are mainly of two types i.e. instrumental (for which instruments are used) and eye or sensory (for which observer uses his senses and are visual) observations.

Table 4 : List of meteorological instruments installed in an agrometeorological observatory along with the measured weather elements and observation timings

S.No	Elements(s)	Instrument(s)	Unit	Observation Frequencey
1.	Rainfall	I. Ordinary gauge II. Self Recording Rain gauge	Millimeters (mm)	I. Twice Daily 8.30 a.m, 2.30 p.m. Continuous (The time of placing the chart 8.30 a.m.)
2.	Air Temperature	I. Dry and Wet Bulb Thermometer II. Min. & Max. Thermometer III. Thermograph	Degree Celsius (°C)	I. Twice daily 0700 IST hrs & 1400 IST II. Twice daily 0700 IST & 1400 IST III. Continuous. The time of removing or placing the chart is 8.30 a.m.
3.	Sun Shine Duration	Sunshine Recorder	Hours (h)	Continuous-The time for placing the chart is before sunrise.
4.	Evaporation	U.S. Class A Open Pan Evaporimeter	Millimeters (mm)	Twice Daily 8.30a.m, 2.30 p.m.
5.	Relative Humidity	I. Dry and wet Bulb Thermometer II. Assmann Psychrometer III. Hygrograph	Per cent (%)	Twice daily 0700 IST & 1400 IST Continuous and the time for placing the chart 8.30 a.m.
6.	Soil Temperature	Soil Thermometer	Degree Celsius (°C)	Twice daily 0700 IST & 1400 IST
7.	Wind Speed	Cup Anemometer	Kilometers per hour (kmph)/ meters per second(mps)/ knots (kts)	Twice daily 0700 IST & 1400 IST
8.	Wind Direction	Wind Vane	Degree clock wise from north on the scale, 00-36 where00 & 36 is the wind from the north, 09 from the east and 00 refers calm;	Twice daily 0700 IST & 1400 IST
9.	Cloud Amount	Oktas (1/8 of the celestial dome)	Visual observed in the observatory	At the time of observation
10.	Dew	Dew Gauge	mm	One time during winter season before sunrise (install after sunset of previous day)

2.3.1. Observation recording timings

Observations in an agro-meteorology observatory are recorded at 0700 and 1400 Indian Standard Time (IST) except for the rain gauge and pan evaporimeter, which observations are recorded at 8.30 and 14.30 IST. For recording accurate weather data, timing of data recording is important therefore, punctuality of observer is a serious consideration in getting quality data. The observation recording time along with weather parameters are given in Table 4.

2.4. General Instructions for Recording Meteorological Observations

- The reading should be written down directly into pocket registers with sharp pencil neatly and immediately after it taken from instrument(s).
- All observations made at a particular hour must be written in the same line of pocket register.
- After taking all the Agrometeorological observations, the meteorological observer should go to office and should properly enter these observations in the CWS forms.
- Observation entered in each page of the pocket register should be copied in ink on corresponding CWS returns.
- Sunshine and dew observations should be written down on daily basis in separate forms.

2.5. Guidelines for Recording Meteorological Observations

- The person recording meteorological observations should go in agro-meteorological observatory in the morning and afternoon hours and first of all wind speed and wind direction are to be recorded form cup anemometer and wind vane respectively. After wind observations, observations of dry bulb temperature, wet bulb temperature, maximum temperature and minimum temperature are taken, and properly entered in pocket register. Soil temperatures have to be recorded from soil thermometer after air temeprature observations. The recording of rainfall data from rain-gauges and evaporation data from pan evaporimeter are to be made at 0830 hours and chart of self recording rain gauge has to be replaced.
- Grass minimum thermometer reading should be taken before sunrise.

- After sunset sunshine card is changed and grass minimum thermometer is exposed.
- In winter months, first of all, recording from dew gauge has to be done before sun rise.
- Observations of anemometer, wind vane, soil thermometers, Max, Min, DB, WB thermometers are recorded at 0700 hrs and 1400 hrs Indian Standard Time (IST).
- Open pan evaporimeter, rain gauge, self recording instruments are recorded and set at 0830 IST.

2.6. Order of Observations

The instrumental observations at the observatory should be read in the following order starting at 10 minutes preceeding the prescribed hours:

1. Wind instruments
2. Thermometers
3. Rain-gauges
4. Barometer

Non instrumental observations *viz*., clouds, visibility should be taken after instrumental observations. All observations should be finished in 10 minutes before the prescribed time of observations.

2.7. Installation of Agro-meteorological Observatory

Agro-meteorological observatories are installed with a very specific purpose of crop-weather relationship studies and therefore, it is installed in a 'site which is fully representative of the crop-soil-climate conditions of that area. Standard arrangement of instrument is strictly adopted as any slight deviation in accuracy of installation of any instrument will affect the weather observations. Thus, setting of agro-meteorological observatory requires care in its installation. IMD is the authorized government agency for authorization of an observatory and generally installation is done under the guidance of an IMD official. The instruments should be procured from a firm approved by IMD and instruments are certified by IMD only.

2.7.1. Selection of site

For obtaining reliable and accurate weather data site selection is very important so that weather data truly represent to the specific area. Following specific points be considered for selecting a site for the meteorological observatory.

- Flat rectangular plot with 55 meters in north-south direction and 36 meters in east-west direction.
- The observatory should be representative of crop-soil and climate condition of area.
- To avoid any obstruction, the site should be away from hill, high buildings, trees, and irrigation channels particularly on the eastern side.
- It should be free from problems of water logging even during heavy rain and easily accessible throughout the year.
- If any necessary building is required, that should be in the North direction so that shadow of building may not fall upon any of instruments.
- The observatory should be located at a site, which serve as a standard.
- It should be located away from steep slopes, water bodies and working engines as this may alter the conditions around the observatory.
- The surface should be cleaned regularly and be covered with thin grass as barren ground causes increased ground radiation.

2.7.2. Lay-out Plan of an Agro-meteorological Observatory

A standard lay-out plan, suggested by IMD is depicted in figure 2.

2.7.3. Precaution in installation of instruments

- The site should be fenced with barbed wire upto a height of 1.5 meter fixed to the iron pole of 2.0 meters height. These iron poles are kept at a distance of 2.5 meter from each other.
- Instruments should be robust, simple and easy to handle and maintain. These should be as per India Meteorological Department (IMD) specification.
- The installation of the instruments should be as per IMD standards.

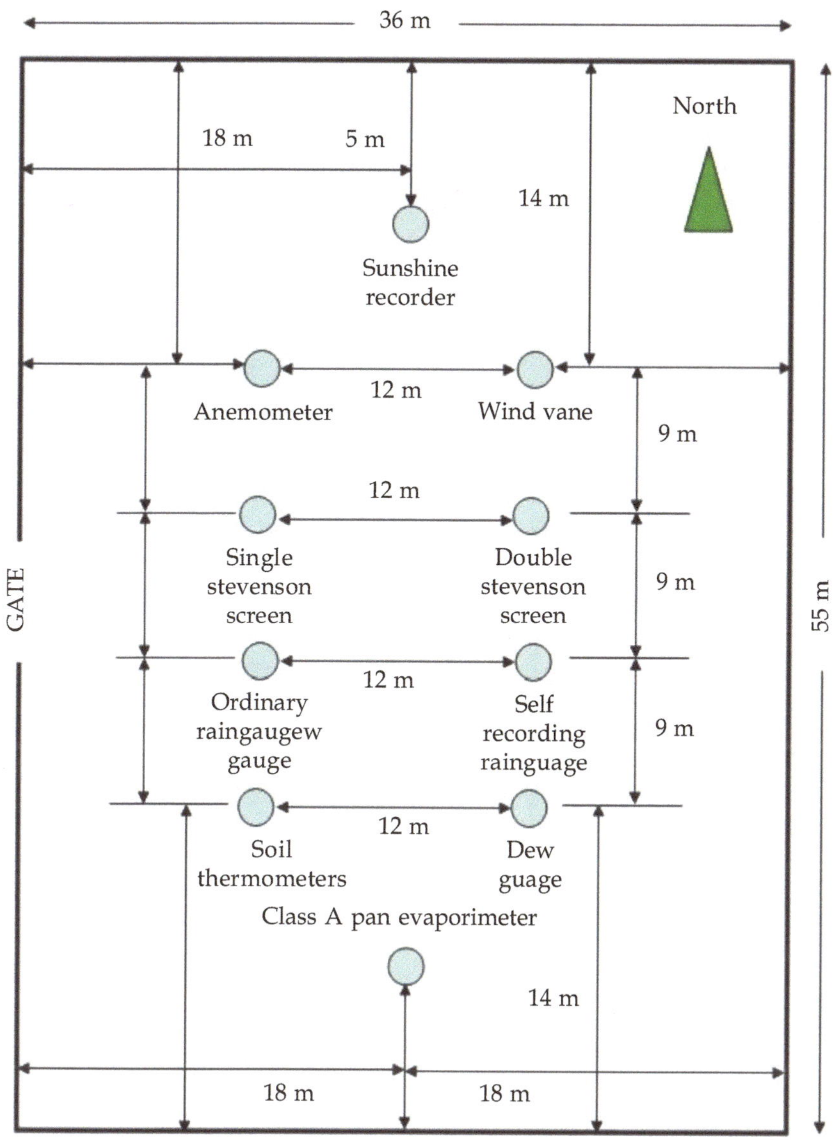

Fig. 2 : Standard layout of Agromet Observatory
PLOT SIZE : 55 m (N-S) x 36 (E-W)

- Tall instruments should be on one side (Northern side) in descending order in the observatory so that they may not affect the observation of the other small instruments.
- The sketch of the layout should be kept as permanent records.

Questions for Study

Q.1. What is the purpose of an Agromet Observatory?

Q.2. Draw a sketch of Agrometeorological Observatory layout.

Q.3. On reaching the observatory which weather variable will be recorded first and then immediately followed by the first?

Q.4. Distinguish between eye-recording and self-recording instruments.

Q.5. Which weather element is most variable?

Q.6. Describe different types of Agromet Observatories.

❑❑❑

Chapter 3

Measurement of Rainfall

Rainfall is precipitation (falling of any type of condensed moisture to the ground) in the form of liquid drops larger than 0.5 mm in diameter falling on the earth surface. Ordinary rain drop size varies from 0.5-4.0 mm in diameter. Rainfall is measured in term of depth of rain water (millimeters) on a solid surface provided that rain water is neither allowed to percolate nor runoff or loss in any form.

3.1. Measuring Instrument

The device used for the measurement of rainfall is called rain gauge. There are two types of raingauges *viz.,* Ordinary raingauge or non recording type and Self-recording raingauge or recording type.

3.1.1. Ordinary raingauge or non recording type

The instrument measures the total amount of rainfall in a stated period (between two observations). It is most commonly used instrument for recording the quantum of rainfall. The standard raingauge is Symon's pattern widely accepted all over the world. Since 1969, on the recommendation of World Meteorological Organization, Fibre Glass Reinforced Polyesters (FRP) rain gauges are used for measurement of rainfall all over the world (Fig. 3).

FRP Rain Gauge

The body (collector and Base) of this rain gauge is made up of Fibre Glass Reinforced Polyesters material, which is more durable and non-rusting under field/open conditions. It consists of collector above a funnel leading into a receiver placed inside a base (Fig 3). Its essential parts are shown in figure 4 and described below:

In ordinary rainguage two types of funnels are used. Wherever rainfall is heavy and frequent, funnel of 200 cm^2 rim area is used.

- *Funnel :* It is placed on a vessel which is made up of FRP material. It is of circular shape with gun metal rim of diameter 112.9 mm and 159.6 cm and cross section area, correspond to 100 and 200 cm^2 respectively.
- *Bottle :* The polythene bottles are of three stages having capacities of 2, 4 and 10 liter of water. It is made up of polythene/plastic and rainfall falling through funnel get collected in the bottle.
- *Collecting jar :* The excess rain water from the bottle is collected in this jar.
- *Measuring cylinder :* It is made up of white clear glass and graduated in teths of millimeter. This cylinder is used for rainfall measurement and having capacity of 20 mm with least count of 0.1 mm.

Fig. 3: Ordinary raingauge

Fig. 4 : Essential parts of Ordinary raingauge

3.1.1.1. *Working principle*

The depth of rainfall falling on the ground is to be measured without any loss either from the evaporation, infiltration or percolation. The rain water entering the rain gauge at the receiving funnel is collected through bottle. The amount of rainfall is measured by measuring glass.

3.1.1.2. *Observation*

Remove funnel, take out polythene bottle, pour collected rainfall into measuring glass graduated in volume units, cc or ml and read the bottom of meniscus. If rain water is more than 20 mm, make suitable installments as required. One division of measure glass is equivalent to 0.2 mm of rain. Generally, total rainfall is reported for the last 24 hours.

The volume of rainwater is calculated by measuring the area of the collector. The area of the collector is estimated by the diameter of rain gauge and its cross section area πr^2.

$$\text{Rainfall} = \frac{\text{Rainwater collected (cubic cm) or ml}}{\text{Area of the collector (sq cm)}}$$

3.1.2. *Self-recording raingauge*

This type of instrument is designed to record the continuous record of the rainfall along with the times of onset and cessation. Thus, it not only records the amount of rainfall but also enables the user to calculate the rainfall intensity. The recording rain gauges are of three types *viz.,* Tipping bucket gauge (Fig. 5). Weighing gauge and Float syphoning gauge type, which is most suitable and accurate instrument for recording the rainfall.

3.1.2.1. *Tipping bucket raingauge*

It is based on the principle of tipping and measures the number of times or frequency that a bucket or pair of buckets fills with water and overbalances. It has a funnel to receive the rainfall and a clock -work system for recording the rainfall. A drum is fixed on the clock work and a graduated chart is fixed on the drum with horizontal and vertical scales. Its essential parts are:

Fig. 5: Outer view of Tipping bucket raingauge

Fig. 6: Inner view of Tipping bucket raingauge

- *Receiver :* It has a cylindrical receiver of 30 cm diameter with a funnel inside. Below this receiver there is a collector.
- *Base :* There are also two types of interchangeable bases.
- *Collector :* It consists a pair of two buckets (Fig. 6). When the rain water is transmitted through funnel into one of the buckets, it is completely filled in 0.25 mm rain. Then it tips and empties into a cylinder placed below, other bucket comes in position, it is filled fully in 0.25 mm and it also tips and empties. This process is repeated.
- *Clock-work system :* The tipping bucket activates electrical circuit and makes marking on the graph paper fixed on the drum. The drum is rotated by clock mechanism. Its sensitivity is 0.10 mm or 0.25 mm. This rain gauge is used in automatic weather station.

3.1.2.2. Float syphoning rain-gauge

The instrument is a float gauge which when filled to a certain level is emptied by syphoning. The instrument has few moving parts and its functioning cannot be easily spoiled by dirt or corrosion.

It is made of FRP (Fibre glass reinforced polysters) body most preferably being non-rusting. The rainwater entering the guage at the top of the cover is led via the funnel to the receiver consisting of a float chamber and a syphone chamber. The main parts of the instrument are Funnel, Float chamber, Strip of bottom of float, Entrance tube, Float, Float rod, Clock drum and Syphon chamber and tube as shown in figure 7.

- *Funnel :* The funnel rim has gun metal edge with diameter 203.2 mm. The rain water collected by the funnel is poured into the float chamber through filter.
- *Float chamber :* It is a cylindrical vessel with plastic float inside. In this chamber a light, hallow float is attached, which has nortical movement with the rise and fall with water level. The upper side of the float is connected with a rod called float rod. The other end of the float rod is attached with a pen arm which moves on a graph paper wrapped around the clock driven rotating drum. A graduated chart calibrated in mm is fixed on the drum with the horizontal and vertical scales and moves with clock work. The horizontal scale indicates the time and vertical scale indicate the amount of rainfall.
- *Syphoning device :* A syphoning device is provided in the float chamber which operates with float rise and its fall. The float when reaches to the highest level it automatically empties the float chamber.

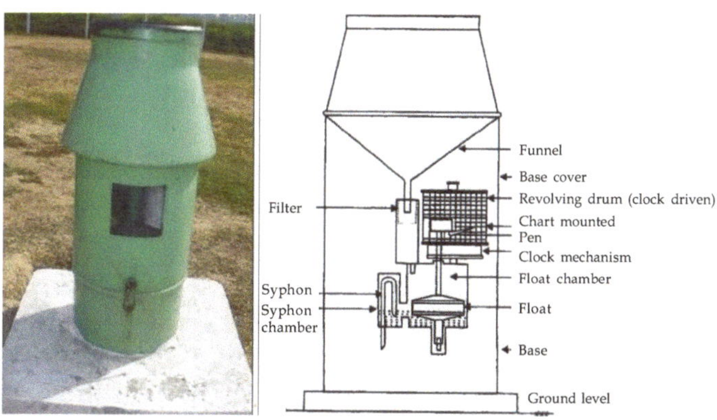

Fig. 7 : Float syphoning rain-gauge and their parts

Now-a-days varies types of floating type recording raingauges are available. Natural Syphon recording rain-gauge is widely used in India.

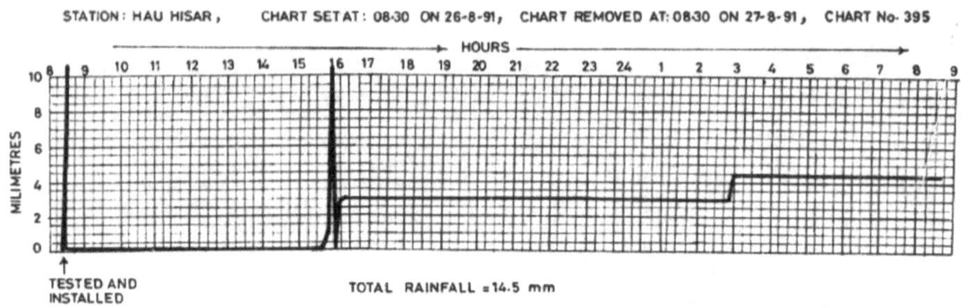

Fig. 8 : Self recording raingauge chart

If there is no rain, the pen traces a horizontal line from where it leaves off rising.

3.1.2.2.1. Working principle

It is based on the principle of syphon. A drum is fixed on the clock work. A float is attached to drum for recording the rainfall (Fig. 7). When the float reaches to the highest level the pen will mark 10 mm on vertical axis and if rain continues syphon is operated automatically and whole chamber is emptied; the rod falls down to zero level and instrument becomes ready to record next rainfall.

3.2. Intensity of Rainfall

Intensity of Rainfall is measured by self recording raingauge. It is the ratio of the amount of rain in a rainy spell to the duration of the rainy spell. It is expressed in cm/hr or mm/hr units.

The low intensity rain provides more time to soil for infiltration and their energy of splash will be low and will not encourage soil erosion. Hence, low intensity rains are more useful for soil moisture. While high intensity rain may cause runoff, soil erosion and may damage the standing crop.

$$\text{Intensity of rainfall (mm / hr)} = \frac{\text{Amount of rainfall}}{\text{Duration of rainfall}}$$

3.3. Installation of Raingauges

3.3.1. Ordinary raingauge or non-recording rain gauges

The gauge is placed on a levelled ground at the standard height of 30 cm above the ground surface. In order to avoid the loss of rain drops due to obstruction, the distance of any object should be at least twice the height of the object above the rim of the raingauge. It is installed in 3′ x 3′ x 3′ (cubic feet) foundation pit size. Fill the lower 1′ with lime concrete and in the center of the pit, construct a brick work structure of 2′x 2′x1′10″. Place the raingauge at the center at the top of the brick work. Put 4″ cement concrete in such a way that 2″ portion of the body of raingauge is inside the ground level and 2″ portion is above the ground and base is embedded so that rim of the gauge is 30 cm above the ground (Fig. 9).

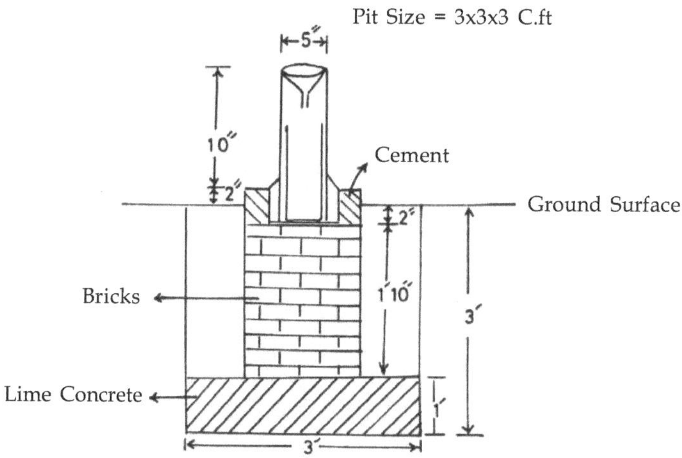

Fig. 9 : Installation of ordinary raingauge

3.3.2. Self-recording raingauge

The gauge is installed on a foundation platform of 3′x3′x3′size as shown in Fig. 10. Fill the 1′ portion with lime concrete. In the center of the pit, construct pillar of 2′x2′x2′ in brick work. Construct the cement concrete of 3″ thickness at the top of the brick work (above ground surface) and insert 9″ long and ¾″ diameter pipe at the base of the concrete with proper slope to drain the water from raingauge (Fig. 10). Care is taken so that the rim of receiving funnel is perfectly horizontal and it should be at a height of 75 cms above the ground surface and fixed on cemented platform. All the foundation work should be done according to the measurements as shown in the figure.

3.4. Time of Observation

Rainfall observation is recorded twice a day at 8.30hrs (morning) and 2.30 hrs (afternoon). Daily observation is total at 0830 a.m. which gives past 24 hrs rainfall record. The rainfall is measured in mm and tenths of mm. For no rains, record 0.0 and for rainfall less than 0.1 mm, record 't' meaning trace.

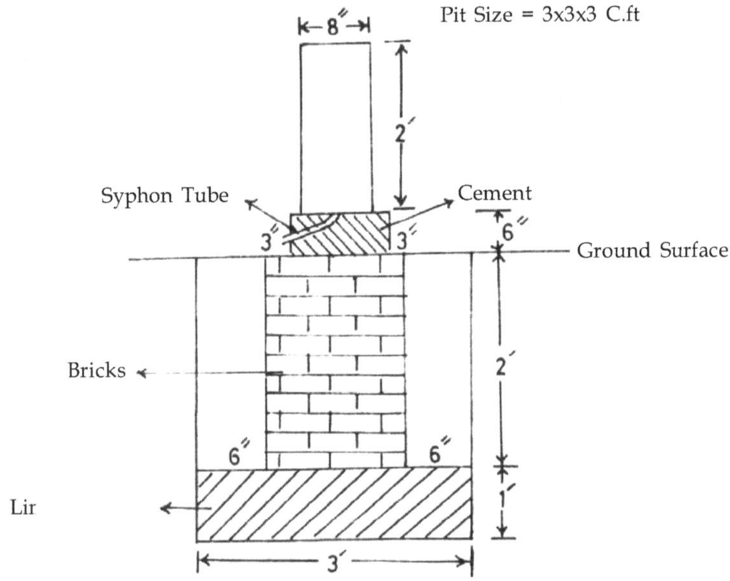

Fig. 10 : Installation of self recording

3.5. Precautions

The rain water should be poured carefully into the rain measuring cylinder taking care to avoid spilling of any drop of water. The measuring cylinder should be placed on a horizontal surface while recording the lower meniscus.

- Measuring should be as per IMD standard.
- The rim of the funnel should be placed horizontally.
- Test the siphon working system on daily basis.
- The recording chart should be changed daily.
- It should be fixed on a base of concrete with proper drainage.

3.6. Plastic Rain Gauge

To measure rainfall now a days plastic rain gauges are used, which are ordinary type of raingauge (Fig. 11). It is very easy to install it and is available in affordable price.

It measures the accumulation (height) of precipitation in an open topped cylindrical container. Operator records the accumulation and empties the cylinder. Normally it has a funnel shaped accumulator at the top to gather precipitation from a greater area than the cylinder area.

Plastic Rain Gauge is made with a weather-resistant plastic gauge with clear, tough butyrate and scale is marked on gauge in metric graduations. Different sizes of plastic raingauges are available in market which can measure from 6" (150 mm) to 11" (280 mm) of rainfall (Table 5). The minimum graduation is 0.01" (or 0.1 mm). Screws are provided to mount the gauge.

Fig. 11 : Graduated Cylinder Rain Gauge

Table 5 : Specifications of different types of plastic raingauges

Type of raingauge	6"	11"
Capacity	6" (150 mm)	11" (280 mm)
Resolution	0.01" (0.1 mm)	0.01" (0.2 mm)
Size	2.5" W x 2.25" D x 13.5" L (64 mm x 57 mm x 343 mm)	4" dia x 14" H (102 mm x 356 mm)
Weight	0.2 kg	0.9 kg

Questions for Study

Q.1. What are the different parts of raingauge?

Q.2. What are the exposure requirements for the installation of a ordinary raingauge?

Q.3. What is the difference between Tipping bucket raingauge and plastic raingauge?

Q.4. How is rainfall measured?

Q.5. Write about the precautions and procedure of rainfall measurement.

Q.6. What is intensity of a rainfall spell and how it is measured?

❑❑❑

Chapter 4

Measurement of Snowfall

Snow is precipitation of solid water, mainly in the form of branched hexagonal crystals or stars. It is "precipitation of white opaque grains of ice". Measurement of snow cover is related to runoff estimation and also important for hilly agriculture.

4.1. Snowfall

It is the amount of fresh snow deposited over a specified period (generally 24 hours). The measurements relate to depth and water content of the snowfall.

4.2. Measuring instruments

Depth of snowfall is measured with help of graduated scale or by snowgauge.

4.2.1. Graduated scale

It directly measures the depth of fresh snowfall. When a fresh snowfall is deposited on open ground surface in the absence of strong winds, than simple graduated scale is used to measure the depth of snowfall (Fig.12). The mean value of several measurements (vertical) gives the depth of snowfall over a surface. Under strong winds conditions large number of vertical observation is required to measure the actual depth of fresh snow.

4.2.1.1. Precaution

- Measurement of old snow should be avoided.
- Place a covering (wood) on the top of the snow surface or paint the top surface of the snow.

4.2.2. Snowgauge

It is a gauge of 20cm in diameter, which is fitted into snow and installed in suitable place for recording the fresh snow depth. The shielded gauge is used to catch the snow in windy conditions (Fig. 13). To avoid the drifting snow catch the gauge is mounted on 3 to 6 meter above the snow surface.

Fig.12 : Graduated scale

Fig.13 : Snowgauge

4.2.2.1. Precaution

- Measurement of old snow should be avoided.
- The gauge should well above the average snow level.
- The site should be carefully selected to reduce the wind effect.

4.3. Water Content of the Snowfall

Water content is an important factor in crop growth and measured with help of snow sampling instruments. Sampling device gives the depth as well as water content of the snow.

4.3.1. Measuring instrument

4.3.1.1. Tubes

Coupled sections of light weight aluminum tubes are inserted vertically through the snow to obtain a sample which is weighed to determine its water equivalent.

4.3.1.2. Isotopes

It is an alternative method to estimate the water content of the fresh snow. In this method radioactive isotope (Cobalt -60) is inserted to the bottom of a snow layer. The isotope transmits gamma rays at intensity proportional to snow water content.

4.3.1.3. Snow pillow

It is flat liquid filled container (pillow), when snow accumulated on the flat surface creates internal pressure and activates electrical contacts (Fig. 14). A solar cell battery, and radio assembly at the remote site transmits water content data to the receiving stations.

4.3.2. Estimation of water content of the snowfall

The water content is approximated by simple relationship between depth of snowfall and water 1 cm of fresh snow = 1 mm of water.

It is also estimated by weighing or by melting of the collected snowfall. Snow collected in snowgauge should be melt immediately after recording & measured by means of an ordinary raingauge measuring glass.

Fig. 14 : Snow pillow

4.3.3. *Precaution*

- Observer should select the representative point during sampling survey.
- The average depth is calculated after taking observation on number of sample point.
- Suitable method is used to obtain the water equivalent.

Questions for Study

Q.1. What is the importance of snowfall measurement?

Q.2. How the water content of a snowfall is measured?

Q.3. What do mean by a snow pillow?

Q.4. How snowfall is measured?

Q.5. Is it possible to measure the snowfall with graduated scale?

❑❑❑

Chapter 5

Measurement of Air Temperature

Temperature influences most plant processes, including photosynthesis, transpiration, respiration, germination, flowering, adaptation and distribution of crops. As temperature increases (up to a point), photosynthesis, transpiration, and respiration increase. When combined with day-length, temperature also affects the change from vegetative (leafy) to reproductive (flowering) growth. Depending on the situation and the specific plant, the effect of temperature can either speed up or slow down this transition.

Temperature is the measure of mean kinetic energy per molecule of the molecules in an object, while the heat is the measure of total kinetic energy of all the molecules of that object. A large object may have a much lower temperature than a small object and may still have greater heat content by virtue of larger number of molecules in it. Thus heat and temperature are not the same.

5.1. Air Temperature

Air temperature is the temperature of the air recorded by the thermometer exposed in a standard type of screen called Stevenson's screen.

5.2. Stevenson's Screen

Stevenson's Screen is used for housing different thermometers and recording instruments for measuring air temperatures. It is also known as shelter screen and provides normal environment for the housing instruments. It is specially designed rectangular wooden box which admits a free movement of air. The box is painted white to reflect the solar radiations. A double roof with intervening of air space prevents direct heating and also from the precipitation to the instruments. The lower height of the box is 4 feet from the ground surface (Fig. 15). The box is always facing north side in Northern Hemisphere and south side in Southern Hemisphere so that only minimum sunlight would enter into the box at the time of taking observations. It has following 3 different sizes.

5.2.1. Single stevenson screen

It is used for housing maximum, minimum, dry bulb and wet bulb thermometers for measuring air temperatures. It's internal dimensions are 56.0 cm length, 31.5 cm width and 41.2 cm height (Fig.15).

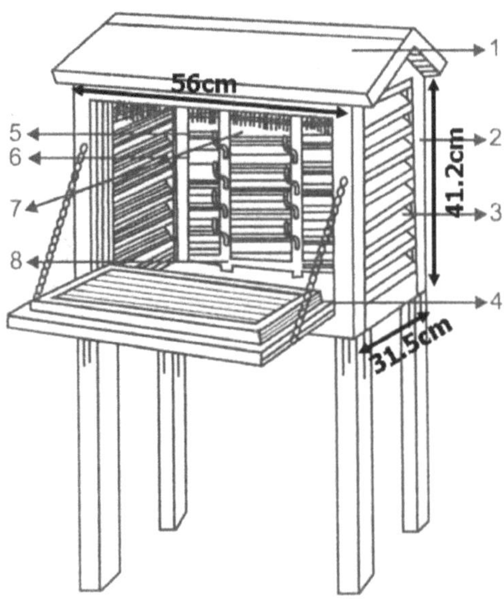

1. Roof
2. Wooden Box
3. Wooden Panes
4. Door
5. Dry Bulb Thermometer
6. Wet Bulb Thermometer
7. Minimum Thermometer
8. Maximum Thermometer

Fig. 15 : Single Stevenson Screen

5.2.2. Double stevenson screen

This screen has double the length of single Stevenson screen and used for housing self recording instruments like thermograph, hygrograph, thermo-hygrograph, barograph etc. The size of this screen is similar to single Stevenson's screen but double in length, size: 112 cm length, 31.5 cm width and 41.2 cm height.

5.2.3. Mini thermometer screen

It is small in size and a portable thermometer screen. It is similar to single Stevenson's screen and has the internal dimensions of the screen are 40.4 cm length, 20.0 cm width and 33.7 cm height.

5.3. Thermometers

5.3.1. Maximum and minimum thermometers

Maximum and minimum thermometers (Fig. 16a, 16b) indicate the highest and lowest air temperature attained during 24 hours since their last setting. Setting is done once a day in the morning for maximum thermometer and in the afternoon hours for minimum thermometer. The maximum thermometer is mercury in glass thermometer with a constriction in the bore between bulb and beginning of the scale. This constriction allows the mercury to move in forward direction with rise of temperature, but does not allow coming back with decrease of temperature.

Once the mercury attains the extreme position with respect to air temperature, it stands at that level in the capillary for recording at the observation hours as the maximum temperature attained since the thermometer was last set. The thermometer is reset after the observations of morning hours by swinging it steadily in a semi-circular arc with the bulb end downwards until the mercury column is reunited and this reset reading should agree with reading of dry bulb thermometer. Range of maximum thermometer lies between -35°C and 55°C and minimum thermometer is between -40°C and 50°C. Reading can be taken with accuracy upto 0.1°C in both the thermometers.

The minimum thermometer is a glass thermometer containing alcohol with a small dumb bell shaped index in the stem. This index is placed in the alcohol with its end initially just touching the meniscus at the end of the column. Surface tension prevents the index breaking through the surface. When the temperature decreases alcohol contracts and the index slides along with it towards the bulb due to force of cohesion (surface tension). The minimum air temperature can be noted on the graduated capillary tube. But when temperature rises before the observation, the liquid moves but the index remain stationary. The farthest end of the index from the bulb indicates the minimum temperature recorded since the thermometer was last set. The index can be reset by tilting the thermometer with the bulb higher than the stem after afternoon reading and this reset reading will be close to the readings of dry bulb thermometer. Both maximum and minimum thermometers are read twice daily at 0700 hrs and 1400 hrs IST.

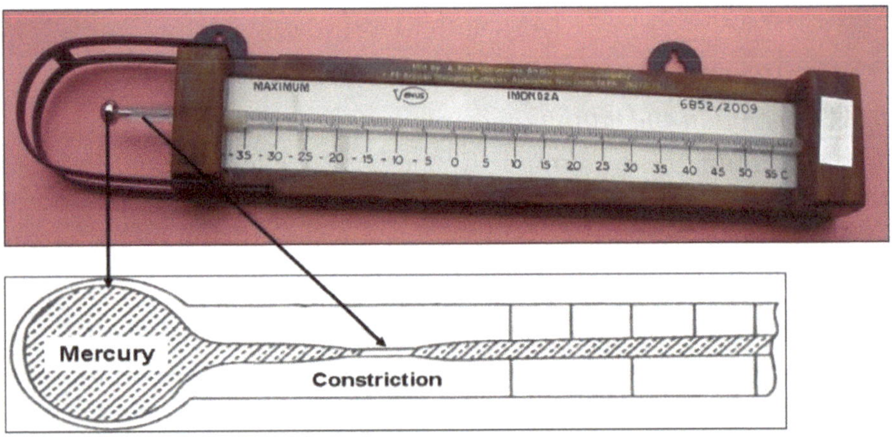

(a) Maximum thermometer

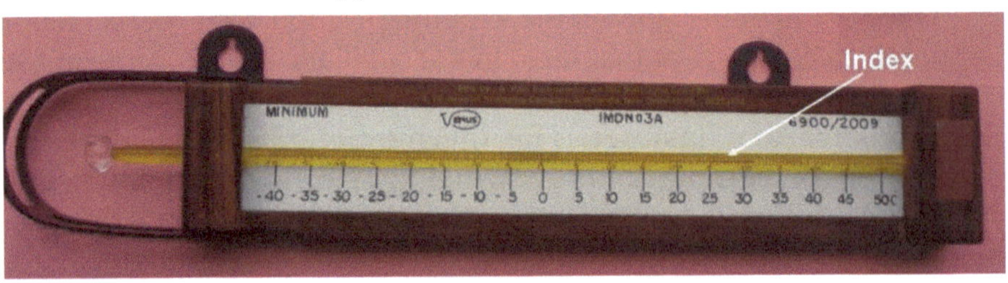

(b) Minimum thermometer

Fig. 16 (a & b)

5.3.2. Dry bulb and wet bulb thermometer

Dry bulb thermometer (Fig. 17) records the instantaneous reading of air temperature, whereas the wet bulb thermometer records air temperature when its mercury bulb is surrounded with water vapours at saturation level at the environmental air temperature and pressure. These thermometers consist of a mercury bulb at one end and connected with a tube which contains a hole in which mercury is filled. The bulb of wet thermometer is covered with a piece of muslin cloth which is kept constantly moist by a wick dipping into distilled water. Cooling of the wet bulb thermometer is proportional to the rate of evaporation of water from its wick. The rate of evaporation in turn depends upon saturation deficit of the surrounding air. When the air is very dry, rapid evaporation of water from the wick cools the wet bulb thermometer. The wet bulb thermometer (Fig.17) thus registers atmospheric temperature reduced by evaporation from a moist surface. Range of dry and wet bulb thermometers lie between -35°C and 55°C and reading can be taken with accuracy upto 0.1°C.

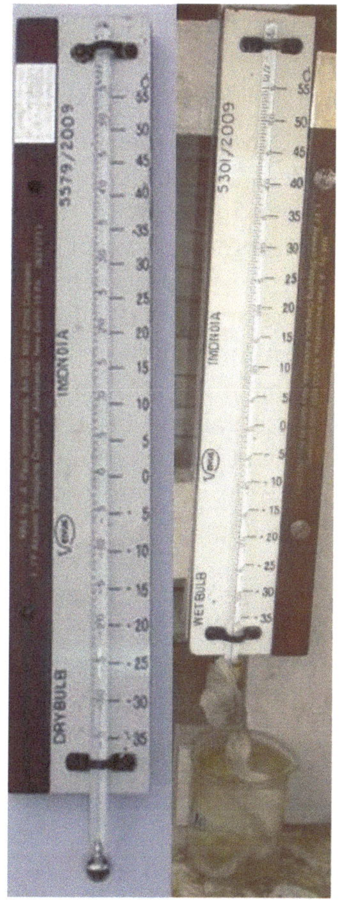

Fig. 17 : Dry bulb and wet bulb thermometers

5.3.3. *Working principle*

All liquids expand and contract with increase and decrease in temperature. Mercury also expands and contract with the temperature of the surrounding air. The change in the mercury level can be recorded as the temperature of the ambient air changes.

Alcohol in the capillary tube expands and contract whenever there is change in the temperature of the surrounding air. Change in the alcohol level is recorded with help of moving metallic index.

1. The freezing point of mercury is -38.3°C.
2. The freezing point of alcohol is -130°C.

5.4. Installation of Single Stevenson's Screen & Thermometers

The foundation pit is 3′ x 2′ x 2.5′. Fill the lower 1.5′ portion with lime concrete. Put a central mark and fix the geographical north with help of magnetic compass. Fix the Stevenson's screen box on the legs of wooden or iron stands, by placing the legs in provided socket. Fix the screen in such a way that the door of the screen should be in north direction. Put 10" cement concrete and after setting put 2" loose soil layer above cement concrete (Fig.18).

The four thermometers are installed inside the "Single Stevenson's Screen". The screen is very useful for exposing liquid-in-glass thermometers to measure air temperatures. The maximum thermometer is kept on the upper two metal brackets on the upper position of the screen and these are positioned so that the thermometer resets at an angle of 2° to the horizontal with the bulb at lower end. The minimum thermometer is laid in position on the lower two brackets on the lower portion of the screen. It also slopes at an angle of 2° to the horizontal and the bulb should be lower than the stem. The dry bulb and wet bulb thermometers are mounted in a vertical position at least 5 1/2" apart from each other and these are fixed in grooves of small brass clips.

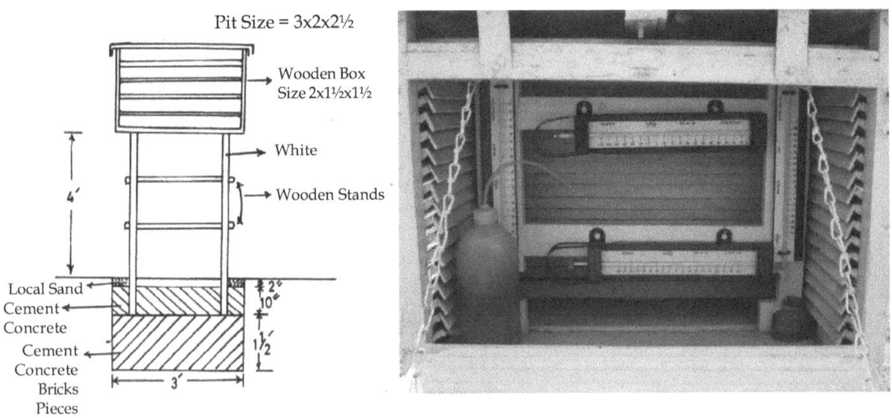

Fig. 18 : Installation of Single Stevenson's Screen and thermometers in wooden box

The installation of Stevenson screen is done on the base which is made as per the measurement and items given in the figure18. The shelter is usually mounted on an open wooden support, the floor of the screen being at 4 ft. (1.22 m) above the ground. The shelter is set up with its door facing north side so that only minimum sunlight would enter while the observer is reading the instruments.

5.5. Observation

- ❒ Open the Stevenson's screen.

- Note the reading of dry bulb thermometer, correct to 0.1°C corresponding to the end of the index away from the bulb.
- Set the maximum thermometer during morning observations (0700 hrs IST) and record the set reading.
- Note the reading of minimum thermometer corresponding to the end of the index away from the bulb.
- Set the minimum thermometer during evening observations (1400 hrs IST) and record the set reading.

5.6. Precautions

- Thermometers should be placed in an adequately ventilated shelter.
- After recording the observation of maximum temperature it should be reset at 7 hrs IST.
- After recording the minimum temperature the thermometer should be tilled and reset the previous position at 2 hrs IST. No jerks should be given.
- Distilled or rain water should be used for wetting the bulb otherwise salt deposition on the bulb does not allow to record the actual temperature.
- The muslin cloth and wick must be kept clean and moist. In case of dust storm, muslin cloth should be replaced.

5.7. Problems and their Rectifications in Mercury (Maximum) and Alcohol (Minimum) Thermometers

5.7.1. Break in mercury column

If break in mercury is found, hold it firmly with bulb down and then swing it till the break in the column removed.

5.7.2. Babbles in alcohol column

Hold the thermometer horizontally. Try to bring the bubble at the starting point of the glass tube near the bulb by slowly tilting it. Hold the bulb in your palm for sometime which may cause warming leading to disappearance of the bubble.

5.7.3. Index enters in the bulb end

Reverse the thermometer slowly and bring back the index in normal position. Then put it in cold water bath for its normalization.

5.7.4. Index gets stuck up

Hold the thermometer vertically and tap it gently. The index should fall in alcohol column. If it does not fall then reverse it slowly and bring back in normal position. When the alcohol column moves to the other end and traps the index in the column. Use the cold water bath to normalize the thermometer.

5.7.5. Deposition of alcohol at the top in thermometer

Tap the end of thermometer gently and than hold the thermometer in cold water, added with ice upto the height of alcohol column keeping the top of thermometer towards sun. Vapour is attracted towards cooler region and the alcohol droplets come down.

5.8. Thermograph

Thermograph is an instrument for obtaining continuous and automatic recording of air temperature (Fig. 19). The thermograph consists of a temperature sensitive bimetallic element which is connected by a system of linkages to a pen recording on a chart fixed on a drum driven by mechanical clock. Two metals having widely different coefficients of thermal expansion like invar and bronze or steel and brass are welded together and rolled in the form of a thin single strip the bimetallic element. It is either bent to form on arc of a circle or would into a close helix of several turns, with the metal having the higher temperature coefficient of expansion on the outside of the helix. One end of arc or helix, is attached to the frame of the instrument, while the other end is fixed to a horizontal spindle to which the pen arm is screwed when the temperature changes, the variations in expansion of two metals takes place with the increase or decrease of curvature of arc and thus helix either coils or uncoils. These variations are transmitted to the pen point where these are recorded on a chart fixed on a revolving clock drum (Fig. 20).

Fig. 19 : Thermograph

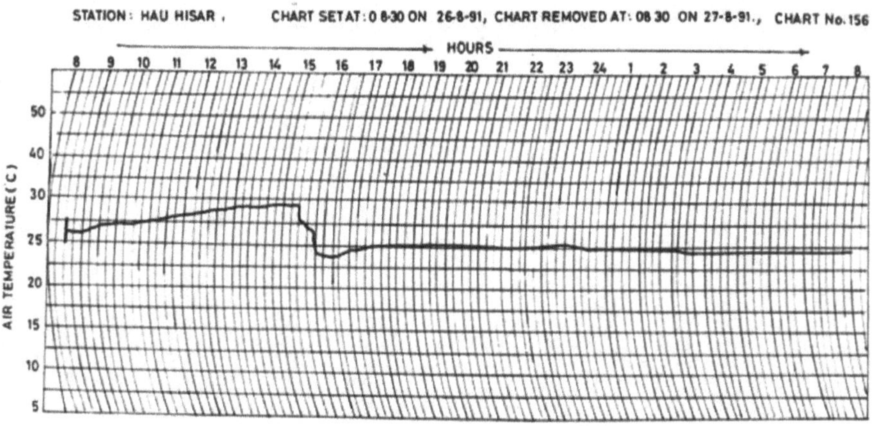

Fig. 20 : Thermograph chart

5.8.1. Precautions

- The instruments for setting of charts with respect to time scale on drum by changing of charts, care of pens, tabulating the charts, winding the clock and general cleanliness should be carefully followed.
- The chart is having horizontal scale in hours and vertical scale in °C for air temperature with a range of 0 to 40°C.
- The drum completes on round in 24 hours with a curve on the graph for a continuous recording of air temperature for 24 hours.

5.9. Other Instruments for Measurements of Air Temperatures

5.9.1. Sling psychrometer or Whirling psychrometer

In the sling or whirling psychrometer, two identical thermometers T1 and T2 mounted on wooden frame, the thermometers are mounted side by side on metal frame, which has a handle enabling it to be whirled. One thermometer is fitted with a sock which is part of a specially woven cotton tube and is moistened.

This can be rotated around vertical axis with the handle. Sling or whirling psychrometer is often insufficiently shielded against radiation (Fig. 21). In this instrument, aspiration is provided by whirling or rotating the thermometers. Rotation should be given so as to obtain a speed of 3 to 5 meter per second. This requires about 4 to 5 revolutions per second.

Fig. 21 : Sling or whirling psychrometer

5.9.1.1. Observation

The whirling psychrometer should be protected from direct sunlight during data recoding time. The effect of radiation is reduced by rotating the instrument in front and windward, and kept in the shade. To obtain the desirable air speed of at least 2.5 meter/sec past the thermometer bulbs and 30 cm long psychrometer should be rotated at a rate of about four revolutions per second. The wet-bulb should be moistened immediately before the observation is started. The psychrometer is whirled for about two minutes, stopped smoothly and read rapidly. The procedure is then repeated. If the readings are not identical the whirling is repeated until two consecutive readings are the same. Care should be taken not to breathe on the thermometers when taking a reading. The measuring steps are:

1. Moisten the wet bulb using minimum amount of water.
2. With back to the sun or in shade, rotate the instrument.
3. Continue rotating, temperature of the wet bulb will start failing and after some times it will not show further lowering of temperature. Take the reading of wet bulb first to an accuracy of 0.1°C.
4. Read the dry bulb thermometer also and record both the readings.

5. Find out RH, Dew point, Vapour pressure. Use special humidity tables specifically worked out for this instrument.
6. Reading should be taken after 15 seconds of whirling. Successive reading should be taken not more than 5 seconds after.

5.9.2 Artificially ventilated psychrometers

Humidity values obtained from psychrometers exposed in a standard screen are subject to uncertainty; due variation in rate of movement of air past the bulb. This uncertainty is generally ignored for ordinary synoptic or climatological purposes. When more accurate values are required, the psychrometer is artificially ventilated at a controlled rate by whirling or by an electric motor or a clockwork motor. Such psychrometers are said to be "aspirated". One of such psychrometer is Assmann Psychrometer (Fig. 22).

It consists two identical mercury in glass thermometers of dry and wet bulb with accuracy of 0.1 °C and range -10 to +60°C which are supported by metal clip. It is clockwise ventilated (aspirated) psychrometer and thermometers are protected from direct radiation by suitable covering with high polished double walled radiation shield. Thermometers are insulated from the frame. There is a clock driven exhaust fan which sucks air inside from the bottom. Air imparts heat to thermometer bulbs and temperatures are recorded. It is used to obtain vertical temperature and humidity profile. The levels of measurement are at surface 1, 2, 4, 8 and 12 feet from ground.

Fig. 22 : Assmann psychrometer

5.9.2.1. Observation

Before taking the measurement, the wet bulb thermometer covered with a wick should be moistened by means of moistening device attached to the instrument. The water absorbed by the wick is sufficient for moistening of the thermometer bulb for approximately 15 minutes when measurements are made in dry air with normal temperature. When temperatures are high more frequent wetting is carried out.

If the natural temperature of the psychrometer before taking a measurement was similar to the temperature of the air, the first utilizable reading can be taken 3 minutes after the starting of the aspiration. Both the thermometers of the psychrometer should be read within effective running time of the aspiration immediately one after the other when steady position is attained.

The following procedure should be followed in making an observation:

1. Moisten the wet bulb;
2. Wind the clockwork motor (or start the electric motor);
3. Wait until the wet bulb reading has become steady (i.e. possibly 2 or 3 minutes);
4. Read the dry bulb thermometer;
5. Read the wet-bulb thermometer; and
6. Check the reading of the dry bulb thermometer.
7. Obtain RH, dew point and vapour pressure using special humidity tables.

Great care should be taken to avoid any influence on the readings by the presence of the observer.

5.9.2.2. Maintenance

The frame, motor housing and radiation shield should be kept well polished using clean chamois leather. The bearing of the aspiration fan can be cleaned, if necessary, by mean of a match stick cut to a point.

5.9.2.3. Precautions

- Use only distilled water for moistening the wet bulb.
- Keep the muslin cap clean and change it when necessary.
- Every month put a drop of clock oil in fan bearings.
- Stable readings of the thermometers should be recorded.
- Keep the mercury bulb of the thermometers on the windward side while recording the observations.

- Place shelter from direct solar radiation and avoid precipitation falling on the dry bulb.

5.10. Other measuring instrument

The air temperatures are measured generally with mercury thermometers but the accuracy of these thermometers is about 0.5°C and also the response time is very large about a minute. Therefore, air temperature are measured with sophisticated instruments with the help of thermistors and thermocouples, the accuracy (0.1°C) and response time (second) of these instruments are faster than mercury thermometers.

5.10.1. Resistance thermometer or thermistors

To measures the change in resistance of an object, specially constructed semiconductor devices called thermistors are the most popular (Fig. 23). Older Resistance Temperature Detectors (RTD) devices use platinum, or nickel alloy (with iron and copper), wound as a fine wire coil. Resistance is changing with change in temperature. To record temperature, resistance response is measured from which gives the air temperature.

5.10.2. Thermocouple thermometer

It measures the voltage produced by the contact of two dissimilar bonded (welded) metal wires (Seebeck effect). Sometimes a pair of wires joined at both ends is used, where one end is held at a reference temperature. The bimetallic junction develops a small voltage that varies with temperature and records the changes in temperature (Fig. 24).

Fig. 23 : Thermistors

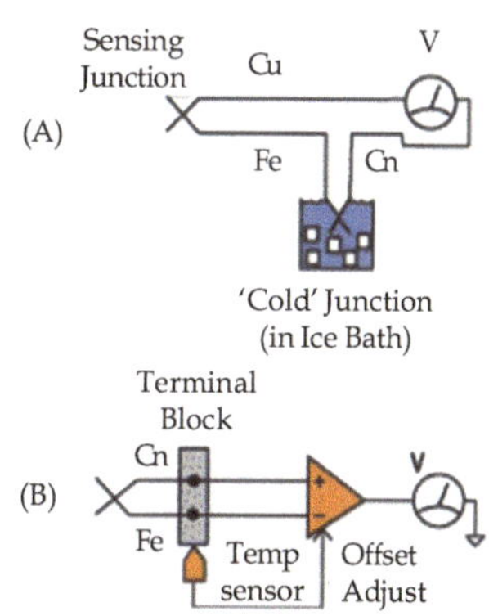

Fig. 24 : Thermocouple

Questions for Study

Q.1. What is Stevenson screen and what are its functions?

Q.2. Why special construction is given to the Stevenson's Screen? Why it is painted white?

Q.3. What is the difference between heat and temperature?

Q.4. How is maximum temperature measured?

Q.5. What is wet bulb temperature and how it is measured?

Q.6. Write about the precautions of air temperature measurement.

Q.7. Why Stevenson screen is always installed facing north in Northern Hemisphere?

Q.8. What is the correct position for reading the thermometer?

Q.9. At what temperature the Celsius and Fahrenheit scales will read same value?

Q.10. Why alcohol is use in minimum thermometer?

Q.11. Why constriction is place in maximum thermometer?

Q.12. Why bulb of wet bulb thermometer is not dipped in ground water?

❑❑❑

Chapter 6

Measurement of Soil Temperature

The surface of the earth gets heated up during the day and gets cooled during the night causing diurnal changes in the top layers of the soil. Soil temperature is an important part of micro-environment, where plant roots grow and capture the water and nutrients and thus, it affects the plant growth and yield. During crop germination also soil temperatures play vital role. Since the heat regimes of these layers are governed by the soil temperature, the measurement of soil temperature becomes extremely important.

6.1. Measuring Instrument

Soil thermometers are used for measuring temperatures of the soil layers at shallow depths which show a marked diurnal variation. Measurements of temperatures at 5, 15 and 30 cms of soil depths are recommended for agromet observatories. The mercury-in-glass thermometers are generally used which are installed on a specially designed stand.

6.2. Soil Temperature Measurement

The soil temperature is measured by soil thermometers. These are mercury-in-glass thermometers of the enclosed scale type. There is a bend of 120° just above the bulb, the rest of the stem being straight. So that when the soil thermometer is installed at a particular depth of the soil, the bulb rests horizontally. In these thermometers there are graduations for every degree centigrade and till graduations begin from the distance of 6.5 cm, 17.5 cm and 35.0 cm from the bulb for the 5, 15 and 30 cms depth soil thermometers, respectively (Fig. 25). To support the thermometers at the correct inclinations, iron stands with sloping sides at 60° to the ground surface are provided. The soil thermometers are installed at a site in the observatory which is sufficiently away from obstructions and is free from water logging during the rainy season.

Fig. 25 : Soil thermometers

6.2.1. Working principle

Mercury expands whenever there is a change in the temperature of the surrounding soil. Change in mercury level can be read on the graduated capillary tube which is equivalent to the temperature of the surroundings soils.

6.2.2. Check distance

The check distance should be checked from time to time i.e. the distance between the ground and the 5 °C mark on the thermometer scale is called check distance. This distance at the time of installation is measured and must be maintained always.

6.2.3. *Observation*

The temperatures are noted by eye observation correct upto 0.1°C. The line joining the eye and the top of the mercury column should be at right angles to the stem of the thermometer. The thermometers should be read from a distance without touching and shadow of the observer should not fall on the thermometers. Take the daily observations and record them in a data sheet.

6.3. Installation of Soil Thermometers

The thermometers are mounted on triangular stands fixed into the ground as shown in figure 26. Stands are made from iron strips of 2.5 cm wide and thick. The slopping side of the stand is inclined at 60° to the ground surface so as to provide support for the correct inclination of thermometers. The stem of the thermometer is held in position by two spring clips.

Soil thermometers are installed inside the observatory enclosure to southern end. The stands are fixed at 45 cm apart with graduated marking facing south. Pits for the thermometers are dug in front of the stands on southern side.

Soil should be removed layer by layer and kept in separate leaps in the order as these are removed. After the thermometers are placed in position, the pits are refilled layer by layer from the soil leaps, mentioned earlier, in the reverse order. The top of the soil should be adjusted so as to remain intact with the undisturbed surface of the plot.

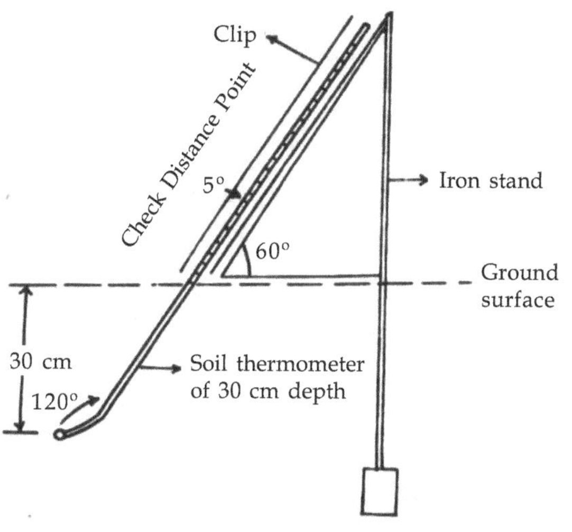

Fig. 26 : Installation of soil thermometer

6.3.1. Precautions

- ❐ The bulb of the thermometer should be kept at the desired depth at an angle of 60° with the horizontal.
- ❐ The tube of thermometer is installed with an iron stand.
- ❐ Parallax error should be avoided during taking the observation.
- ❐ At the time of observations sit sufficiently away from the soil thermometer to avoid shadow near the thermometer.
- ❐ No rain water be allowed to stand near soil thermometers.
- ❐ Read correct upto 0.1 °C soil temperature at 0700 and 1400 hrs (IST) in order of 5, 15 and 30 cms soil depths, respectively.
- ❐ The plot should be free from any vegetation by removing weed grasses.
- ❐ In dry weather, some soils may develop cracks and development of cracks may be delayed by raking the soil surface gently when hot season is approaching.

Questions for Study

Q.1. How is soil temperature measured?

Q.2. Write the precautions of soil temperature measurement.

Q.3. Why soil temperature face south in North hemispheres and vice versa?

Q.4. Why soil temperature is higher at surface during summer?

Q.5. Why soil temperature at 30 cm depth is higher than 5 cm depth during winter?

Q.6. What is check distance?

❐❐❐

Chapter 7

Measurement of Grass Minimum Temperature

It is the actual minimum temperature experienced by the plants near ground surface. The grass minimum temperature is lower than the air temperature at 4 ft level. It is the temperature recorded in open air ground on short turf, with the bulb of the thermometer just in contact with the tips of the blades of grass. It is also described as the temperature at 5.0 cm (2 inches) above ground. It is used during October to March period only when frost occurrence is expected. The readings of the grass minimum thermometer indicate the possibility of ground frosts, which is important information for farmers, gardeners and growers.

7.1. Grass Minimum Thermometer

It is a sheathed minimum thermometer and used to measure temperature of vegetation or grass at a height between 2.5 and 5.0 cm above the soil surface (Fig. 27).

7.2. Measuring Instrument

The grass minimum temperature is measured by a minimum thermometer. The bulb is linked-shaped and provides a larger surface for exposure than a spherical bulb. As with all minimum thermometers the actual thermometer is held inside a glass jacket and the contraction of the alcohol column overnight causes a small indicator within the stem of the thermometer to descend and record the lowest temperature.

Fig. 27 : Grass minimum thermometer

7.2.1. Working principle

Alcohol in the capillary tube expands and contract whenever there is change in the temperature of the surrounding of the grasses on the earth surface.

7.3. Installation Grass Minimum Thermometer

Mark two plots each of size 3′x 2′ side by side and prepare raised beds of 5 cm height. Plant grass on beds and allow it to grow between 2.5 and 5 cm height, trim it if height is more. Place the thermometer horizontally on two small Y shape wooden bases in such a position that thermometer just touches the tips of the grass. When one bed is irrigated or the grass is required to be trimmed, the thermometer is placed in the second bed.

7.4. Observation

The thermometer is set and exposed 2.5 to 5.0 cm above ground surface on its support over the grass plot in the evening and its reading is noted before sunrise, next mornings. After the reading is over, the thermometer should be removed and kept in the screen to some shady place (indoors) to avoid direct exposure to solar radiation, bulb downwards, until the evening when it should be reset.

7.4.1. Observation timing

It is kept after sunset for observation and readings are recorded just before sunrise.

7.5. Precautions

After reading in the morning, instrument should be removed and kept to some shady place (indoors) to avoid direct exposure to solar radiation; otherwise the column broken.

Questions for Study

Q.1. How grass temperature is recorded?

Q.2. Which thermometer will explain better occurrence of frost in the winter season?

Q.3. Why grass minimum thermometer reading is lower than minimum thermometer of Stevenson's screen?

Q.4. What height of grass is needed to record the grass minimum temperature?

Q.5. Why mercury is not used in grass minimum thermometer?

❑❑❑

Chapter 8

Measurement of Humidity

The presence of moisture in atmosphere is termed as humidity and plays an important role in formation of weather system and plant as well as insect-pest growth and development. An abundance of moisture results in a rich natural flora and makes possible a wide choice of crops. Deficiency of moisture, on the other hand, permits only narrow range of crops. Humidity also affects the water requirement and the rate of transpiration. Reduction in transpiration reduces the translocation of food material and uptake of nutrients. High humidity with high temperature favours the outbreak of pest and disease. Moisture stress condition reduces the cell size and shape and ultimately plant growth and yield is reduced. High soil moisture and humidity condition cause failure in establishing stand of legumes and pulses. Excess of soil moisture in earlier growing season may result in more vegetative growth and delayed flowering.

8.1. Measures of Humidity

The following terminologies have been used to express the presence of water vapour in air.

8.1.1. *Absolute humidity (AH)*

It may be defined as the mass of water vapour per unit volume of air or it is the ratio of mass of water vapour to the volume of air. The unit is g/cm^3.

$$\text{Absolute humidity (AH)} = \frac{\text{Mass of water vapour (mv)}}{\text{Volume of the air (V)}}$$

8.1.2. *Specific humidity (SH)*

It is defined as the mass of water vapour per unit mass of moist air. Specific humidity is expressed in grams of water vapour in kilograms of air (g/kg). It decreases with increase in latitude. So, it is higher at equator and lower at poles.

$$\text{Specific humidity (SH)} = \frac{\text{Mass of water vapour (mv)}}{\text{Mass of moist air (mv + ma)}}$$

where, ma = Mass of dry air

8.1.3. *Relative humidity (RH)*

It is defined as ratio of the mass of water vapour present in air at a temperature to the mass of maximum water vapour that air can held at the same temperature. It may also be defined as the ratio of actual vapour pressure to the saturated water vapour pressure.

$$\text{Relative humidity (RH)} = \frac{\text{Mass of water vapour in a given volume of air at a temperature}}{\text{Mass of maximum water vapour that air can held in the same volume and at same temperature}} \times 100$$

OR

$$\text{Relative humidity (RH)} = \frac{\text{AVP}}{\text{SVP}} \times 100$$

where, AVP = Actual vapour pressure and SVP = Saturated vapour pressure

Another measure of humidity is the dew point temperature which is the temperature at which air would become saturated if cooled at constant pressure without addition or removal of water vapour. Thus, the actual vapour pressure is equal to the saturated vapour pressure at the dew point temperature. The closer the dew point to air temperature, the nearer is the air to the saturated condition.

8.2. Measuring Instrument

The humidity of air is measured with the help of dry and wet bulb thermometers (discussed in chapter-5) or by hygrograph. The difference between dry bulb and wet bulb readings at a given time and place provides the humidity of the air and dew point temperature. Normally, a large difference indicates a low relative humidity and a small difference a high relative humidity. When dry bulb and wet bulb readings are same, the air attains saturation and the relative humidity is 100 per cent. Actual value of relative humidity is obtained from hygrometric table corresponding to the reading of dry and wet bulb temperatures (Table 7).

8.2.1. Hair hygrograph

It is an instrument used for continuous recording of relative humidity of the atmosphere. It is based on the principle that the length of a human hair freed from fat, varies considerably with the change in relative humidity (Fig. 28). Human hairs are most sensitive and therefore are widely used in hygrographs since eighteenth century. The length of hairs increases by about 2 to 2½ per cent, when the relative humidity is altered from 0 to 100 percent. Although, this overall extension varies from hair to hair, but there is fairly constant relation between relative humidity and elongation of hairs expressed as fraction of overall change in length.

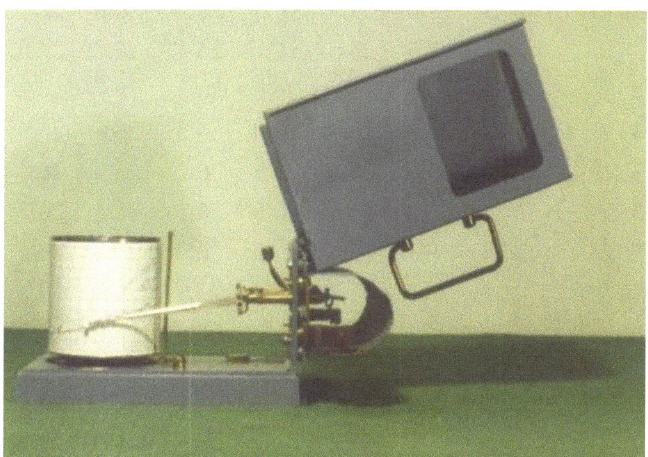

Fig. 28 : Hair hygrograph

Table 6 : Dew point values at different dry and wet temperature values in ° C and tenths.

Dry bulb °C	Dew point °C Wet bulb °C									
	28.0	**28.2**	**28.4**	**28.6**	**28.8**	**29.0**	**29.2**	**29.4**	**29.6**	**29.8**
32.0	26.4	26.7	27.0	27.3	27.6	27.9	28.2	28.5	28.8	29.0
32.2	26.4	26.7	27.0	27.2	27.5	27.8	28.1	28.4	28.7	28.9
32.4	26.3	26.6	26.9	27.2	27.5	27.8	28.0	28.3	28.6	28.9
32.6	26.2	26.5	26.8	27.1	27.4	27.7	28.0	28.2	28.5	28.8
32.8	26.1	26.4	26.7	27.0	27.3	27.6	27.9	28.2	28.5	28.7
33.0	26.0	26.3	26.6	26.9	27.2	27.5	27.8	28.1	28.4	28.7
33.2	26.0	26.3	26.6	26.9	27.1	27.4	27.7	28.0	28.3	28.6
33.4	25.9	26.2	26.5	26.8	27.1	27.4	27.7	27.9	28.2	28.5
33.6	25.8	26.1	26.4	26.7	27.0	27.3	27.6	27.9	28.2	28.5
33.8	25.7	26.0	26.3	26.6	26.9	27.2	27.5	27.8	28.1	28.4
34.0	25.6	25.9	26.2	26.5	26.8	27.1	27.4	27.7	28.0	28.3
34.2	25.5	25.8	26.1	26.4	26.8	27.1	27.4	27.6	27.9	28.2
34.4	25.4	25.8	26.1	26.4	26.7	27.0	27.3	27.6	27.9	28.2
34.6	25.4	25.7	26.0	26.3	26.6	26.9	27.2	27.5	27.8	28.1
34.8	25.3	25.6	25.9	26.2	26.5	26.8	27.1	27.4	27.7	28.0
35.0	25.2	25.5	25.8	26.1	26.4	26.7	27.0	27.3	27.6	27.9
35.2	25.1	25.4	25.7	26.0	26.3	26.7	27.0	27.3	27.6	27.9
35.4	25.0	25.3	25.6	26.0	26.3	26.6	26.9	27.2	27.5	27.8
35.6	24.9	25.2	25.6	25.9	26.2	26.5	26.8	27.1	27.4	27.7
35.8	24.8	25.2	25.5	25.8	26.1	26.4	26.7	27.0	27.3	27.6
36.0	24.7	25.1	25.4	25.7	26.0	26.3	26.6	26.9	27.2	27.6
36.2	24.7	25.0	25.3	25.6	25.9	26.3	26.6	26.9	27.2	27.5
36.4	24.6	24.9	25.2	25.5	25.9	26.2	26.5	26.8	27.1	27.4
36.6	24.5	24.8	25.1	25.4	25.8	26.1	26.4	26.7	27.0	27.3
36.8	24.4	24.7	25.0	25.4	25.7	26.0	26.3	26.6	26.9	27.2
37.0	24.3	24.6	25.0	25.3	25.6	25.9	26.2	26.5	26.9	27.1
37.2	24.2	24.5	24.9	25.2	25.5	25.8	26.2	26.5	26.8	27.1
37.4	24.1	24.5	24.8	25.1	25.4	25.7	26.1	26.4	26.7	27.0
37.6	24.0	24.4	24.7	25.0	25.3	25.7	26.0	26.3	26.6	26.9
37.8	23.9	24.3	24.6	24.9	25.3	25.6	25.9	26.2	26.5	26.8

8.2.1.1. Working principle

The hygrograph depends upon the contraction and expansion of human hair with variation in relative humidity. The changes in the length of hairs are fed to lever mechanism. Longer arm of the lever is connected with pen and ink which makes marking on the graph paper. These changes are transferred on the chart by a pen. The rotating drum with the chart makes a complete circle within 24 hours and the chart has a range of relative humidity from 0-100% vertically with each division equal to 2% and the number of used chart is 398 (Fig. 29).

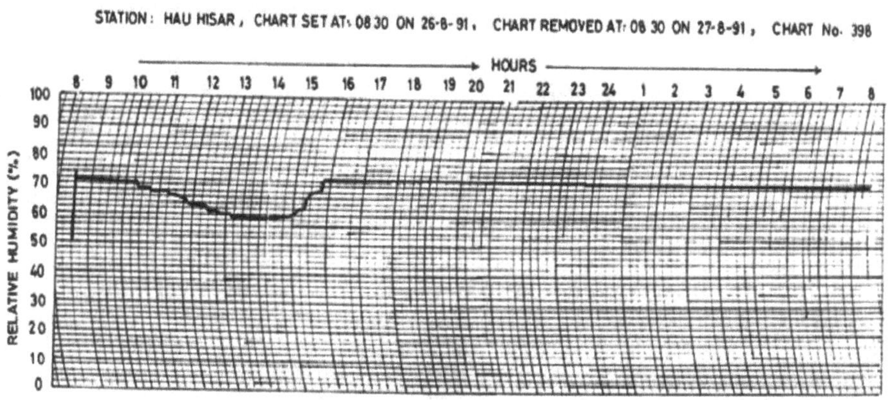

Fig. 29 : Hair hygrograph chart

8.2.1.2. Observations

- Open the Stevenson screen.
- Note the reading of dry bulb thermometer correct to 0.1°C.
- Note the reading of wet bulb thermometer correct to 0.1°C.
- Read from Hygrometric Tables dew point temperature and relative humidity values corresponding to observed dry bulb and wet bulb temperature readings after carrying out necessary interpolation.
- Compute vapour pressure from the dew point temperature obtained, by using the saturation vapour pressure Table.
- Compute the saturation vapour pressure corresponding to the dry bulb temperature from the same Table.
- Calculate the saturation vapour pressure deficit.

8.2.1.3. Precautions

- Instrument is being exposed in double Stevenson's screen.
- The hair needs regularly and thoroughly washing with distilled water. It should not be strained in any way.
- Accurate setting of time scale is very important.
- Hair should not be touched by fingers.

8.3. Calculation

8.3.1. Calculation of actual vapour pressure (AVP)

After taking readings of both thermometers the dew point, vapour pressure and relative humidity may be evaluated with the help of psychrometric tables (Saturation vapour pressure (Table 8).

The psychrometric equation is; $e_a = e_s - r\,(T_d - T_w)$

where e_a = actual vapour pressure, e_s is saturated vapour pressure at wet bulb. r is psychrometric constants (0.5 mm of Hg/°C or 0.67 mb/°C)

AVP (e) = $e'_w - [0.000660(1 + 0.00115\, t'_w)] \cdot p \cdot (t_d - t_w)$

Where,	td	=	dry bulb or air temperature
	t′ w	=	wet bulb temperature
	p	=	barometric pressure
	e	=	actual vapour pressure,
	e′	=	SVP at (t'_w) wet bulb temp.

8.3.2. Calculation of dew point temperature and relative humidity from wet bulb and dry bulb temperatures

From the dry bulb and wet bulb temperature readings, the dew point temperature and relative humidity can be obtained by reference to Tables 6 & 7. Dew point temperature and relative humidity corresponding to specify values of dry and wet bulb temperatures are given in Tables 6 & 7 at intervals of 0.2°C. While using the Tables, interpolation to the nearest 0.1 °C has to be done, wherever necessary. The following example would illustrate the procedure.

Example : Dry Bulb temperature	=	34.5°C
Wet Bulb temperature	=	29.7°C

Table 7 : Relative humidity at different dry and wet bulb temperature values in ° C and tenths.

Dry bulb °C	Relative humidity (%) Wet bulb °C									
	28.0	**28.2**	**28.4**	**28.6**	**28.8**	**29.0**	**29.2**	**29.4**	**29.6**	**29.8**
32.0	73	74	75	76	78	79	80	82	83	84
32.2	71	73	74	75	77	78	79	80	82	83
32.4	70	71	73	74	75	77	78	79	81	82
32.6	69	70	72	73	74	75	77	78	79	81
32.8	68	69	70	72	73	74	75	77	78	79
33.0	67	68	69	70	72	73	74	75	77	78
33.2	66	67	68	69	71	72	73	74	76	77
33.4	65	66	67	68	70	71	72	73	74	76
33.6	64	65	66	67	69	70	71	72	73	75
33.8	63	64	65	66	67	69	70	71	72	73
34.0	62	63	64	65	66	67	69	70	71	72
34.2	61	62	63	64	65	66	67	69	70	72
34.4	60	61	62	63	64	65	67	68	69	70
34.6	59	60	61	62	63	64	65	67	68	69
34.8	58	59	60	61	62	63	65	66	67	68
35.0	57	58	59	60	61	62	63	65	66	67
35.2	56	57	58	59	60	61	62	64	65	66
35.4	55	56	57	58	59	60	61	63	64	65
35.6	54	55	56	57	58	59	61	62	63	64
35.8	53	54	55	56	58	59	60	61	62	63
36.0	53	53	54	55	57	58	59	60	61	62
36.2	52	53	54	55	56	57	58	59	60	61
36.4	51	52	53	54	55	56	57	58	59	60
36.6	50	51	52	53	54	55	56	57	58	59
36.8	49	50	51	52	53	54	55	56	57	58
37.0	48	49	50	51	52	53	54	55	56	57
37.2	48	49	49	51	51	53	53	55	55	57
37.4	47	48	49	50	51	52	53	54	55	56
37.6	46	47	48	49	50	51	52	53	54	55
37.8	45	46	47	48	49	50	51	52	53	54

Table 8 : Saturation vapour pressure in mm of Hg in terms of mean air temperature in ° C and tenths.

Temp. °C	0.0	0.1	0.2	0.3	0.4	0.5	0.6	0.7	0.8	0.9
5	6.5	6.6	6.6	6.7	6.7	6.8	6.8	6.9	6.9	7.0
6	7.0	7.1	7.1	7.2	7.2	7.3	7.3	7.4	7.4	7.5
7	7.5	7.6	7.6	7.7	7.7	7.8	7.8	7.9	7.9	8.0
8	8.1	8.1	8.2	8.2	8.3	8.3	8.4	8.4	8.5	8.5
9	8.6	8.7	8.7	8.8	8.9	8.9	9.0	9.0	9.1	9.1
10	9.2	9.3	9.3	9.4	9.5.	9.5	9.6	9.7	9.7	9.8
11	9.8	9.9	10.0	10.0	10.1	10.2	10.2	10.3	10.4	10.5
12	10.5	10.6	10.7	10.7	10.8	10.9	10.9	11.0	11.1	11.2
13	11.2	11.3	11.4	11.5	11.5	11.6	11.7	11.8	11.8	11.9
14	12.0	12.1	12.1	12.2	12.3	12.4	12.5	12.5	12.6	12.7
15	12.8	12.9	13.0	13.0	13.1	13.2	13.3	13.4	13.5	13.5
16	13.6	13.7	13.8	13.9	14.0	14.1	14.2	14.3	14.3	14.4
17	14.5	14.6	14.7	14.8	14.9	15.0	15.1	15.2	15.3	15.4
18	15.5	15.6	15.7	15.8	15.9	16.0	16.1	16.2	16.3	16.4
19	16.5	16.6	16.7	16.8	16.9	17.0	17.1	17.2	17.3	17.4
20	17.5	17.7	17.8	17.9	18.0	18.1	18.2	18.3	18.4	18.5
21	18.7	18.8	18.9	19.0	19.1	19.2	19.3	19.5	19.6	19.7
22	19.8	19.9	20.1	20.2	20.3	20.4	20.6	20.7	20.8	20.9
23	21.1	21.2	21.3	21.5	21.6	21.7	21.9	22.0	22.1	22.3
24	22.4	22.5	22.7	22.8	22.9	23.1	23.2	23.3	23.5	23.6
25	23.8	23.9	24.1	24.2	24.3	24.5	24.6	24.8	24.9	25.1
26	25.2	25.4	25.5	25.7	25.8	26.0	26.1	26.3	26.4	26.6
27	26.7	26.9	27.1	27.2	27.4	27.5	27.7	27.9	28.0	28.2
28	28.4	28.5	28.7	28.9	29.0	29.2	29.3	29.5	29.7	29.9
29	30.1	30.2	30.4	30.6	30.7	30.9	31.1	31.3	31.5	31.7
30	31.8	32.0	32.2	32.4	32.6	32.8	32.9	33.1	33.2	33.5
31	33.7	33.9	34.1	34.3	34.5	34.7	34.9	35.1	35.3	35.5
32	35.7	35.9	36.1	36.3	36.5	36.7	36.9	37.1	37.3	37.5
33	37.7	37.9	38.2	38.4	38.6	38.8	39.0	39.3	39.5	39.7
34	44.6	44.8	45.1	45.3	45.6	45.8	46.1	46.3	46.6	46.8
35	47.1	47.3	47.6	47.9	48.1	48.4	48.6	48.9	49.2	49.4
36	44.6	44.8	45.1	45.3	45.6	45.8	46.1	46.3	46.6	46.8
37	47.1	47.3	47.6	47.9	48.1	48.4	48.6	48.9	49.2	49.4
38	49.7	50.0	50.2	50.5	50.8	51.1	51.3	51.6	51.9	52.2
39	52.5	52.7	53.0	53.3	53.6	53.9	54.2	54.5	54.7	55.0
40	55.3	55.6	55.9	56.2	56.5	56.8	57.1	57.4	57.7	58.1

(a) Computation of relative humidity :

Referring to Table 7, the relative humidity values of the corresponding temperatures are as follows:

		Wet Bulb	
		29.6° C	29.8" C
Dry Bulb	34.4°C	69	70
	34.6°C	68	69

For this, firstly we have to interpolate for dry bulb temperature as follows:

		Wet Bulb	
		29.6 °C	29.8 °C
Dry Bulb	34.5 °C	68.5	69.5

So for wet bulb temperature of 29.7°C relative humidity is 69%

(b) Computation of dew point temperature :

The values from the tables are as follows:

		Wet Bulb	
		29.6° C	29.8°C
Dry Bulb	34.4°C	27.9	28.2
	34.6°C	27.8	28.1

Interpolating for dry bulb temperatures, we get:

		Wet Bulb	
		29.6° C	29.8°C
Dry Bulb	34.5°C	27.9	28.1

If any digit is followed by 0.5, it should be rounded off to the next higher digit only when the digit is even, and when the digit is odd, it remains the same. *For instance,* 8.5 should be rounded off as 9 and 9.5 also as 9. This procedure is followed in order to avoid systematic over estimation during rounding off.

So, final interpolated value of dew point temperature is 28.0°C for the wet bulb temperature of 29.7°C.

8.3.3. Computation of vapour pressure and saturation vapour pressure deficit

Dew point temperature should be calculated from the Table 6 following the above mentioned procedure. Since the vapour pressure is the saturation vapour pressure at the dew point temperature, the vapour pressure can be computed by noting from the saturation vapour pressure table (Table 8), the value corresponding to the dew point temperature.

For instance, in the previous example dew point temperature is 28.0°C. By referring to the corresponding values in saturation vapour pressure table, we get vapour pressure as 28.4 mm of mercury.

Saturation vapour pressure deficit is defined as the difference between the saturation vapour pressure and actual vapour pressure of the air. In the above example, the saturation vapour pressure deficit can be computed as follows:

Saturation vapour pressure = 45.8 mm of mercury
(Obtained by referring to the saturation vapour pressure
Table 8 entry corresponding to dry bulb temperature of 34.5°C).

Actual vapour pressure = 28.4 mm of mercury.
So, saturation vapour pressure deficit = 45.8 - 28.4
= 17.4 mm of mercury.

8.4. Solved Exercises

Exercise # 1 : Calculate vapour pressure deficit and relative humidity if saturation vapour pressure is 15.0 mm of Hg and actual vapour pressure is 12.0 mm of Hg?

Solution

$$\text{Relative humidity} = \frac{\text{AVP}}{\text{SVP}} \times 100$$

AVP = 12.0 mm of Hg

SVP = 15.0 mm of Hg

$$\text{Relative humidity} = \frac{12.0}{15.0} \times 100 = 80\%$$

Vapour pressure deficit = AVP – SVP
= 15.0 – 12.0
= 3.0 mm of Hg

Exercise # 2 : Calculate relative humidity and specific humidity if the mass of water vapour is 22g in a one kg sample of air at 25 °C and the mass of maximum water vapour in the given sample can held at same temperature is 30g.

Solution

1) $\text{Relative humidity} = \dfrac{22}{30} \times 100$

$= 73\%$

2) $\text{Specific humidity initially at } 25\ ^\circ\text{C} = \dfrac{22\text{g}}{1\text{ kg}} = 22\text{g/kg}$

3) $\text{Specific humidity at saturation level at } 25\ ^\circ\text{C} = \dfrac{30\text{g}}{1\text{ kg}} = 30\text{g/kg}$

Exercise # 3 : Calculate the absolute humidity (AH) if the mass of water vapour (mv) is 20 gm in 1 m^3 volume of air (v).

Solution

$$\text{Absolute humidity (AH)} = \frac{\text{Mass of water vapour (mv)}}{\text{Volume of air (v)}}$$

Mv = 20g

V = 1 m^3

$$\text{Absolute humidity (AH)} = \frac{20\text{g}}{1\text{ m}^3} = 20\text{g/ m}^3$$

Questions for Study

Q.1. What do you understand by humidity of atmosphere?

Q.2. What are the precautions for taking dry and wet bulb reading?

Q.3. What is hygrograph and how it works?

Q.4. How vapour pressure and saturation vapour pressure deficit are computed?

Q.5. How you estimate the relative humidity, if you have dry and wet bulb temperature reading?

Q.6. What are the different measures of humidity?

Q.7. Why human hair is used in hair hygrograph?

Q.8. What is the general problem in self-recording instruments?

□□□

Chapter 9

Measurement of Wind

Wind influences the plant life both physiologically and mechanically. Wind affects plant directly by increasing the transpiration and intake of CO_2. Thus raising the supply of CO_2 to the plants and thereby increase in photosynthesis. However, the increase in photosynthesis is again upto a certain wind speed beyond which its rate become constant. The warm and dry winds increase evaporative loss and cause moisture stress. Lodging is one of severe injury caused by strong winds which causes great yield loss depending upon the stage of crop. This injury is common in paddy, maize, wheat, sorghum and sugarcane. Strong winds damage fruit trees extensively by breaking leaves and twigs and causing fruit drops.

Wind is the air in horizontal motion caused due to differences in atmospheric pressure. Wind has to be specified by its direction and speed. Winds are named after direction from which it comes. It blows from north-east to south-east, it is named as north east wind. Windward refers to the direction from which the wind comes and leeward refers to the direction in which the wind blows. The movement of wind is almost horizontal and the vertical component is very small, being about 1/ 100th of the horizontal component. So by wind, we generally mean only the horizontal component of the wind. Winds transport heat, moisture and other properties from one place to another place in the atmosphere.

9.1. Motion of Wind

Winds are generated due to differential pressure distributed over the earth and in atmosphere. Due to differential heating of earth surface, a low pressure is developed over warmer place and high pressure is developed over colder place. The air moves horizontally from high-pressure area to low pressure area along the pressure gradient as shown in figure 30. There are two types of motion/movement of air. The horizontal motion/movement is know as wind and the vertical motion is know as current.

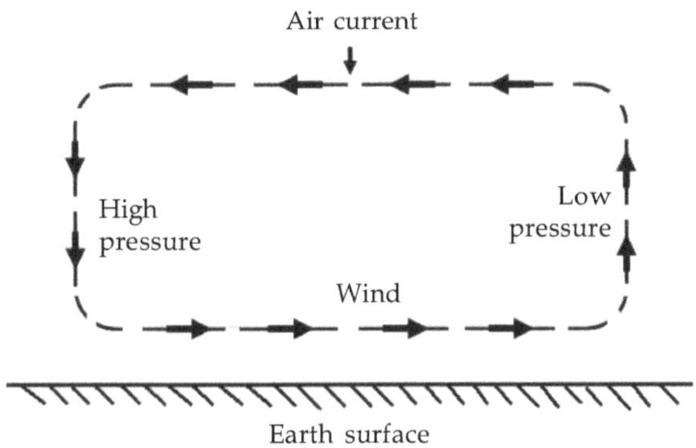

Fig. 30 : Motion of wind

9.2. Measuring Instruments

The Robinson's cup anemometer is used for measurement of wind speed and wind vane is use for determination of wind direction.

9.2.1. Robinson's cup anemometer

The instrument has three cups which are related by winds and its rotation is measured with the help of a gear system. The cups are in semi-conical shape which are attached to a central spider by means of three arms (Fig. 31). The spider is attached to a vertical steel spindle and the top of spindle moves in a small ball bearing while the bottom rests on a thrust bearing. The lower tip of spindle is enclosed inside a bushing which allows free rotation of the spindle. The rotation of the upright spindle is transferred by means of a worm and gear to a counter having bold figures with a range from 00000 to 99999. The four white figures to the left give whole number in kilometers and the last figure in red, gives tenths of a kilometer.

The wind speed is measured in kmph or mph. Meteorologists generally use the wind measurement units either in knots or km (1 knot = 1.853 kmph = 1.52 mph).

Fig. 31 : Robinson's cup anemometer

9.2.1.1. Working principle

The force, due to wind on the concave side of cup is greater than that on a convex side in a similar position and this difference in forces rotates the cup wheel. There is an attachment of gear wheel with a counter which provides wind run directly in kilometers.

9.2.1.2. Measurement of wind speed

To obtain wind speed in kilometers over a given time interval, the counter is read at the beginning and end of a time interval and the difference is computed. The average wind speed during this given period is obtained by dividing the differences for the counter readings by time interval.

9.2.1.2.1. Determination of instantaneous speed at the hour of observation

The anemometer does not respond to wind speed below 1.5 to 2.0 knots but rotates at 2-5 knots. To determine wind speed (V) at the time of observation, take two successive observations of counter at an interval of 3 minutes. Subtract the first reading (a) from the second reading (b) and multiply the difference by 20, if the anemometer is calibrated in kilometres. Thus,

Wind speed (V) = 20 (b-a) Kmh^{-1}

Surface wind speed for synoptic purposes is reported in knots and is obtained by multiplying Kmh^{-1} with 0.54.

Wind speed (V) = 0.54 [20 (b-a)] Knots

This gives the average wind during the 3 minutes of observation.

Example : If the first anemometer reading is 3954.6 and the second reading is 3955.4, the wind speed is given by:

Wind speed = 20 x (3955.4 - 3954.6) = 20 x 0.8 = 16.0 kmph.

9.2.1.2.2. Calculation of mean daily wind speed

The mean daily wind speed is calculated at the 1 day of observation *viz.,* 7 hrs IST at the agromet observatories. The anemometer reading of 7 hrs IST of the previous day is subtracted from 7 hrs. IST reading of the current day and the difference is divided by 24 so that mean daily wind speed in kmph can be obtained. Thus, mean daily wind speed on a particular date corresponds to the 24 hour period ending at 0700 hours IST time of that date. The following example will illustrate this procedure.

Example : If the anemometer reading at 1 day of 17th Nov., 2010 is 2878.3 and anemometer reading at 1 day of 16th Nov., 2010 is 2594.7, the wind speed is given by:

Anemometer reading at 1 day of 17th Nov., 2010 = 2878.3

Anemometer reading at 1 day of 16th Nov., 2010 = 2594.7

Difference = 283.6

Mean daily wind speed on 17.6.82 = 283.6/24 = 11.8 kmph.

9.2.1.2.3. Estimation of wind speed (Indirect method)

In the absence of equipment for measuring wind speed, observation may be made by estimation. Estimates are based on the effect of the wind on movable objects. Beaufort scale (Annexure- IX) may be found useful for estimating wind speed.

8.2.3. Wind vane

The wind vane is an instrument designed for the determination of wind direction. The wind direction is always taken as the direction from which the wind is blowing. This instrument measures instantaneous wind direction.

The vane consists of two flat plates joined together at an angle. The rotating arm is mounted by means of a ball bearing and brass sleeve on wind vane stand. The brass sleeve is fixed by means of two screws on iron rod of the stand. Below the rotating arm, eight fixed direction arms are fixed to a brass bore for indicating main and intermediate directions (Fig. 32a).

9.2.3.1. Working principle

It is a balanced lever which turns freely about a vertical axis. One end of the lever exposes a broad surface to the wind, while the other end is narrow and points to the direction from which wind blows.

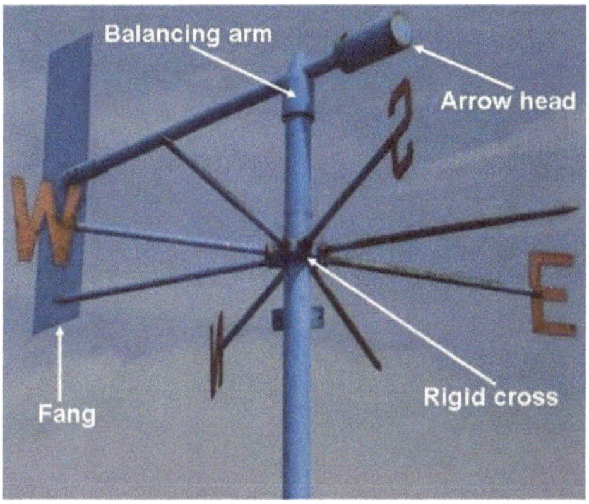

Fig. 32 a : Wind vane

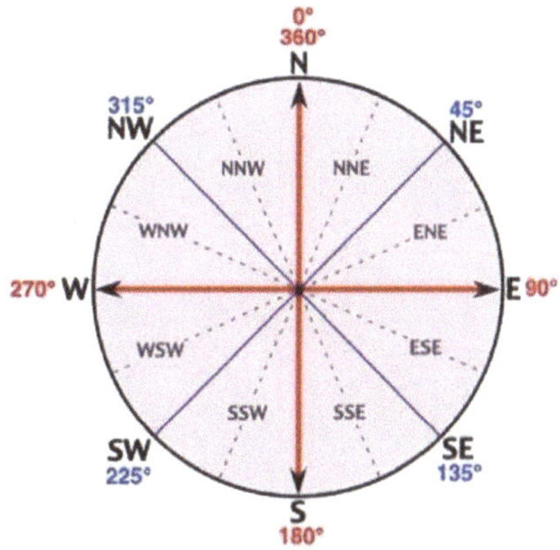

Fig. 32 b : Wind compass showing the wind directions with angle

9.2.3.2. Observation

Wind direction is expressed in degrees, measured clockwise from the north. For reports purpose, it is expressed on the scale from 0°-360° and the total directions as well as code numbers of different directions are indicated in figure 32 b.

9.2.3.3. Measurement of wind direction

The wind direction is read by noting the direction to which the arrow head points. The wind vane should be watched for a few minutes so that the mean direction can be identified. The direction should be read from the wind vane to the nearest of the sixteen points of compass. For example if the arrow head is pointing towards the middle of the region between North (N), and North-west (NW), the direction should be read as North-north-west (NNW).

9.3. Installation of Cup Anemometer and Wind Vane

The pit size for both these instrument should be 2 x 2 x 3 ft^3 and the arrow of wind vane and cups of anemometer also should have an equal height of 3 meter (10 feet) from the ground surface. All the foundation work and the height of wooden post in both cases should be done according to diagram specifications shown in figure 33. The north direction indicator of the wind vane should be set to the true geographical north, which is generally 3 degree east or west to the magnetic north of a place.

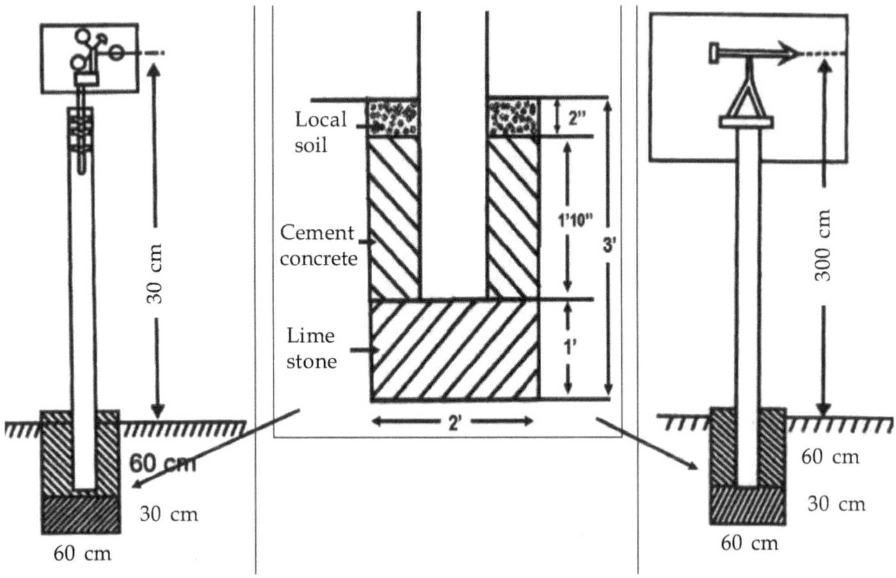

Fig. 33 : Installation of cup anemometer and wind vane

9.4. Time of Observation

The reading of the cup anemometer is taken at 0700 (IST) and 1400 (IST) daily. To take the wind direction observation watch the wind vane for few minutes, after that take the mean direction of the arrow of the vane or take the direction which is shown frequently by the arrow during the time period of the observation. The absence of appreciable motion in air is termed as 'calm' and cup of anemometer will not rotate in this situation. In this situation, no attempt should be made to determine the wind direction from the wind vane.

9.5. Precautions

- The observer should stand as close as possible to the base of support of wind vane and note carefully the direction of the arrow point i.e. the direction from which wind is blowing.
- The observer should also see cup anemometer whether the cups are rotating or not.
- The wind vane should be watched for a few minutes to obtain mean direction of the wind.
- Before taking the reading it should be ensured that the wind vane moves freely.

- The direction should be read from wind vane to the nearest of sixteen points of compass.
- The minimum exposure criterion for the wind instrument is that any obstruction should be away from wind instrument by least 10 times the height of obstruction.
- Lubrication by clock oil should be done every fortnight.
- The height of cups of cup anemometer and arrow level of wind vane should be same at the level for which wind velocity is recorded.

Questions for Study

Q.1. What is wind?

Q.2. Write the importance of wind speed and wind direction in agriculture.

Q.3. How is wind direction measured at the time of observation?

Q.4. How is wind speed measured at the time of observation?

Q.5. What are precautions taken for wind speed and direction calculated?

❑❑❑

Chapter 10

Measurement of Duration of Bright Sunshine

Sunshine means incoming solar radiation, which consists of a bundle of rays of different wavelengths. At the top of the atmosphere, the amount of radiant energy received is found to be constant (1.94 cal/cm^2/min). The receipt of the radiant energy, which is received on the earth and on crop surface, is influence by the cloudy and sunny or clear sky conditions. The brightness hours of the sunshine are an important parameter, which not only influence microclimate of a place but also the photosynthesis of the crop. The radiant energy of intensity of 0.17 cal/cm^2/min or 120 W/m^2 and above is known to specify the bright sunshine.

10.1. Measuring Instrument

The bright sunshine hours are measured with the help of Campbell-Stokes type sunshine recorder. On any particular day, the time and duration can be determined when the sky was covered with or without clouds.

10.1.1. Description of campbell-stokes sunshine recorder

Campbell-Stokes type sunshine recorder consists of a glass sphere of about 102 mm in diameter and a bowl supported by a semi circular brass bar. The focal length of the glass sphere is 74.9 mm. The instrument is mounted on a marble base of 20 cm^2 area. The bowl is made up of gun metal and has a central line engraved transversely across its inner-surface to facilitate its adjustment (Fig. 34). The inner surface of bowl has three sets of grooves for three types of cards. The accuracy of the instrument is 0.1 hour.

Fig. 34 : Campbell-Stoke sunshine recorder

The overlapping pairs of grooves are provided in the bowl to take cards suitable during different period of year.

The semi-circular brass bar of nearly rectangular cross section attached symmetrically to the back of the bowl and concentric with it, forms the main support for the glass sphere. The glass sphere is held at two ends of a diameter by means of brass screws. The brass bar has a scale of latitude (+45° to -5°) and the latitude scale is adjusted according to the latitude of a particular place.

10.1.2. Sunshine cards

There are three sets of cards to suit different time period. These cards introduced in the three different grooves slot. The cards are made from good quality paste board, 0.4 mm thick, coloured with a matt finish blue ink for easy distinguishable of burns on the cards. They are treated to char rather than burn so that the trace produced by strong sunlight does not spread markedly. The long curved cards fit into the bottom set of grooves and the straight cards into the middle set of grooves and the short cards fit into upper set of the groove.

a) Long curved cards

These cards are used in summer season when the days are longer than night during 13th April to 31st August. These cards are introduced through the bottom slot in the concave plate (Fig. 35c).

b) Short curved cards

These cards are used in winter season when nights are longer and days shorter during a period 13th October to end of February. These are introduced at the top slot (Fig. 35a).

c) Straight cards

These cards are used in equinoctial periods when the days and nights are equal for two periods. First from 1st march to 12th April and second from 1st Sept. to 12th October (Fig. 35b).

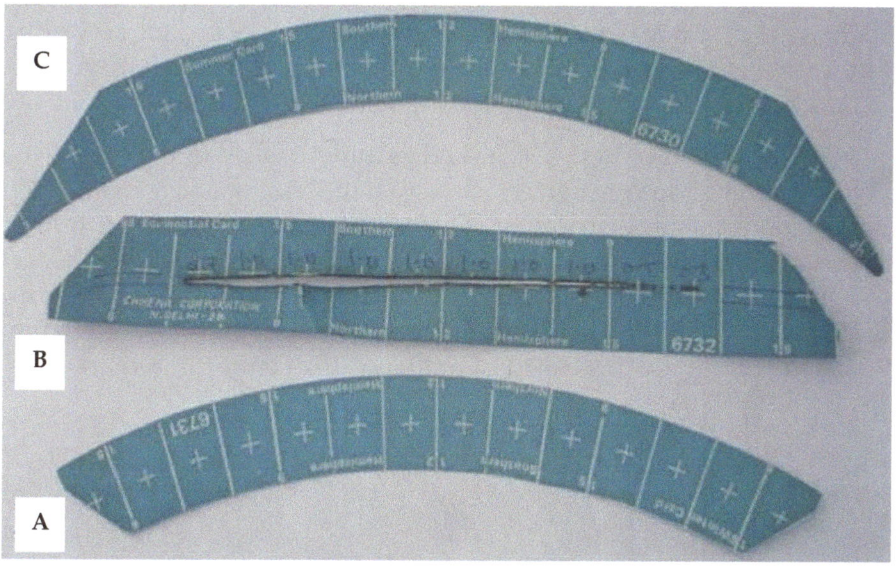

Fig. 35 : Three sets of cards for different seasons, **(A)** Short curved card for winter season, **(B)** Straight card for periods when the days and nights are equal, **(C)** Long curved cards for summer season.

10.1.3. Working principle

The sun rays are focused through a glass sphere to an intense spot which moves with the sun movement across the sky. When sky is clear and the radiant energy of intensity of 0.17 cal/cm^2/min or 120 W/m^2 and above, the intense spot burns a card which is placed at the focal length of the glass sphere. The length of the burns is proportional to the duration of bright sunshine hours.

10.1.4. Time of observation

The recording card proper to the time of the year is inserted every day after sunset and is changed at the same time every day. In rainy season, fresh card can be inserted before sunrise. The card is taken out from the sunshine recorder daily on 7.0 hrs IST. The length of trace gives the duration of bright sunshine hours. When the sun is obscured by clouds, no burn will be produced and indicates cloudiness.

10.2. Installation of Sunshine Recorder

The sunshine recorder is exposed at a height of 3 meter (10 feet) at a place where there are no obstacles to obstruct the sun's rays at any time of the day throughout the year. It is installed in the north side of the observatory on a pillar dimension of 45lenth x 45width x 300height cm. All masonry and brick structural work is done according to the measurements as shown in the diagram. The foundation work would be done in a pit size of 4 x 3 x 4 cubic ft. (Fig. 36).

The sunshine recorder is placed symmetrically on the top of the brick pillar after getting correct level with the help of spirit level which is provided on the instrument. The adjustment for exact latitude of observing station is done by setting the inclination of the brass bar on which latitude scale is provided. The instrument is set in north-south direction with the help of a magnetic compass. The instrument is slightly adjusted till the burn is exactly parallel to the central line marked on the card.

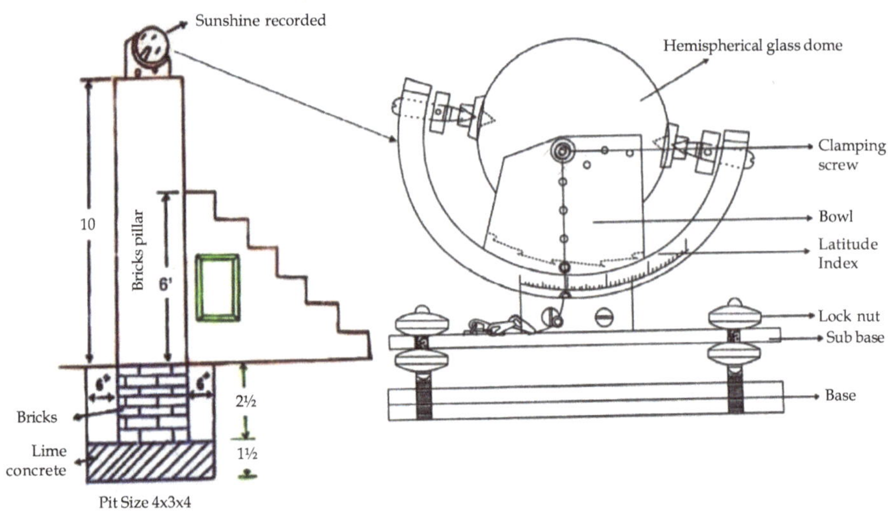

Fig. 36 : Installation of sunshine recorder and Sunshine recorder with different parts

10.2.1. Geographical alignment

The instrument is fixed in North-South direction with the help of magnetic compass. To get the burn parallel with the card, move the instrument till the burn on the card is exactly parallel to the center of the card.

10.2.2. Latitude adjustment

To set the exact latitude of the observing station is made by setting the inclination of the brass bar, sphere and bowl with the help of latitude scale, index and clamping screw provided in the instrument.

10.3. Precautions

- The cards should be changed after sunset or before sunrise each day.
- Date along with station name should be written on back side.
- Perfect horizontal base, latitude adjustment and north-south alignment should be properly done in setting the sunshine recorder.
- Use proper card for appropriate date.
- Remove immediately any deposit such as dew, frost, snow, bird dropping from the glass sphere.
- If the trace of card burn is not parallel to the central line of the card, carry out adjustments of the recorder.

Questions for Study

Q.1. What do understand by bright sunshine?

Q.2. Write working principle of Campbell-Stokes type sunshine recorder.

Q.3. How cloud duration is estimated?

Q.4. What types of cards are used for estimation of bright sunshine?

Q.5. What types of adjustment are essential at the time of installation of sunshine recorder?

❑❑❑

Chapter 11

Measurement of Cloudiness

Cloud is a visible aggregate of minute water droplets and/or ice particles, or both in the air usually above the general ground level. Clouds are the most important form of suspended water droplets caused by condensation. There is no precipitation without clouds.

11.1. Measuring Instrument

The cloudy hours are measured with the help of eye or sensory parts of the body. No particular instrument is required for cloud amount or cloud cover measurement in the sky.

11.2. Cloudiness Observation

Cloud amount estimation covers total amount of cloud present in the sky at the time of observations.

11.2.1. Estimation of cloud amount

Cloudiness is recorded with the help of visual observations at observation hours in terms of 'Oktas' or eights of the sky (Fig. 37) i.e. how many parts of the sky is covered with clouds if the sky is divided into 8 equal parts. Unit for measurement of cloud cover is okta or percentage, which represents 1/8th part of the sky .The cloudy hours in a day, is estimated with the help of sunshine hours data as

$$C = 1-(n/N),$$

where 'C' is cloudiness expressed as fraction between 0 and 1,

n is actual sunshine hours and N is the length of the day i. e. duration between sunrise and sunset hours of the day.

Cloud amount recorded in okta or percentage is reported with the coding convention given in Table 9.

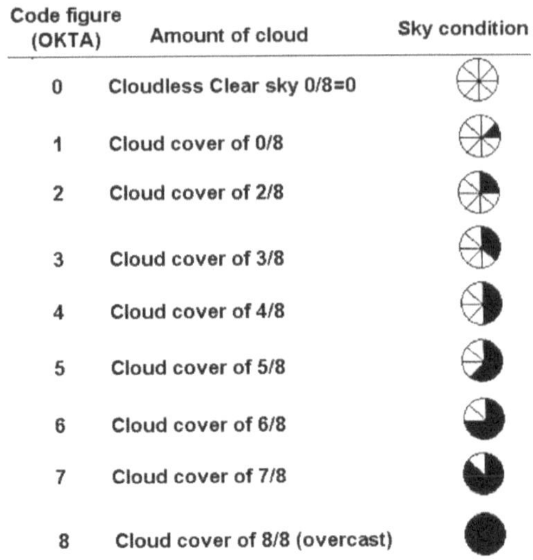

Fig. 37 : Cloud amount and sky conditions

Table 9 : Cloud amount code and their reporting

Cloud Amount		Meaning
Okta	**%**	
0	0	Clear sky/No cloud
1	12.5	Mainly clear sky
2	25.0	Mainly clear sky
3	37.5	Partly cloudy
4	50.0	Partly cloudy
5	62.5	Mostly cloudy/Generally cloudy
6	75.0	Mostly cloudy/Generally cloudy
7	89.5	Almost cloudy/Generally cloudy
8	100	Overcast Sky/total cloud cover
9		Cloud amount can not be estimated due to fog , dust etc or Rain

11.3. Cloud height

The distance between ground and base of cloud is called cloud height.

11.3.1. Cloud height measurement

The measurement of the height of a cloud is done by using a device which is called ceilometer (Fig. 38). It is photoelectric instrument for ascertaining the cloud height. It is a device using a laser or other light source to determine the height of a cloud base.

The Laser Ceilometer measures cloud height and thickness, in addition to vertical visibility, detecting up to four cloud layers simultaneously to a distance of 30,000 vertical feet. A laser pulse is emitted into the atmosphere and backscatter analyzed. Using the speed of light, the altitude of each cloud base and top is determined. Due to poorly defined borders or a sparse composition, some clouds are much more difficult to measure. Depending on the current and historical sky conditions, an adaptive algorithm determines the number of returns needed to maintain accuracy. Accurate measurement of cloud height and thickness in all weather conditions, including heavy precipitation and low clouds, can cause serious errors.

Fig. 38 : Laser Ceilometer

Questions for Study

Q.1. What is cloudiness and how it is estimated?

Q.2. How cloud height is measured?

Q.3. What is different between partly cloudy and generally cloudy sky conditions?

Q.4. What is the meaning of 65 per cent cloudy conditions?

Q.5. What type of instrument is used to measure the cloud amount in sky?

❑❑❑

Chapter 12

Measurement of Solar Radiation and Albedo

The solar energy is the main source of energy for the processes occurring at the earth surface and in its atmosphere. Crop production, is in fact an exploitation of solar radiation and plants are the converters of solar energy into a useful form of energy which is biomass and sustain life providing food, fodder, fiber, fuel, fruits etc. Sun energy fulfills two essential needs of the plants i.e. (i) light for photosynthesis including many other photo processes and functions of the plants, (ii) thermal conditions required for the normal physiological functions of the plant. Thus, the crop production and agriculture is basically dependent on quality and quantity of radiation.

The amount of energy received from other celestial bodies is negligible in comparison to the solar energy. The energy in the form of radiation that strikes the earth is called insolation or the solar radiation that comes to the earth is called as incoming solar radiation and the energy in the form of radiation that leaves the earth surface is called out going radiation. Solar radiation emitted within 0.15-4.0μ wavelength, in general is called short-wave radiation. Earth emits radiation in the wavelength 4.0-100μ and is called long-wave radiation. Of the total radiation emitted by the sun in the entire electromagnetic spectrum, earth receives radiation in 290-4000 nm wave length ranges.

Solar radiation is the main form of energy for all life on earth. While the wave length ranges 350-800 nm are useful for all biological activities on the earth's surface, the wave length ranges below 350 nm are harmful to life. The components of radiation are defined below:

12.1. Different Components of Radiation

12.1.1. Direct radiation

The part of energy which reaches the earth directly is called direct radiation.

12.1.2. Diffuse and scatter radiation or sky radiation

The radiation after scattering by air molecules, dust and water droplets in the atmosphere, that reaches the earth's surface is known as diffuse sky radiation and has the same wave length as the direct radiation.

12.1.3. Total or global radiation

The sum of direct and diffused radiations is called total or global radiation.

12.1.4. Net radiation

It is the difference between the incoming and outgoing components of both short and long wave radiation fluxes.

12.2. Albedo

It is defined as the ratio of reflected shortwave radiation to the incidence shortwave radiation on a surface and expressed in percentage. The albedo of surface is the percentage of the incident radiation reflected by it, is also known as reflection coefficient. The reflection coefficient in case of solar radiation is termed as albedo. Albedo of earth as a whole is also known as planetary albedo or global albedo. Its value is 34 per cent, but it varies with latitude, season and nature of surface. The mean albedo of various surfaces is given in Annexure-V.

$$\text{Albedo} = \frac{\text{Reflected solar radiation by a surface}}{\text{Total solar radiation incident on the surface}} \times 100$$

12.3. Measuring Instrument

All the radiation instruments are broadly classified into three groups:

1. Pyrheliometers,
2. Pyranometers and
3. Pyrradiometers

Direct irradiance at right angles to the solar beam is measured with the 'pyrheliometer' and total solar radiation (direct and diffuse) on a horizontal surface by a 'pyranometer' commonly called 'solarimeter'. Pyrradiometers measure the radiant energy in 0.15 to 4.0 mm in shortwave and 3 to 80 m in longwave bands. When the pyranometer is used in inverted position to measure the reflected solar radiation to characterize the earth and vegetation surfaces called albedo meter.

A net radiometer measures the difference between the incoming and outgoing components of both short and long wave radiation fluxes. Other, less common instruments are the 'pyrgeometer' measuring long wave irradiance on a horizontal surface. For measuring photosynthetically active radiation (PAR) and near infra-red (NIR) radiation, spectral radiometers or band pass solarimeters are used based on photoelectric devices.

These instruments usually employ a thermopile (a series of very closely spaced differential thermocouples) exposed either at the end of heavily insulated narrow tube or under a protective glass hemisphere.

Pyrheliometers are primary standards used for calibration. Almost all the instruments - pyrheliometers, pyranometers, pyrradiometers and net radiometers are calibrated with reference to a standard pyrheliometer and need periodic checking.

The Global radiation values were computed from sunshine hours using Angstrom formula:

$R_s = \{ a + b^* (n/N)\} Ra$

Where,

R_s = Global solar radiation at the surface ($MJ/m^2/day$)

Ra = Extraterrestrial radiation ($MJ/m^2/day$: Annexure-XI)

n = Actual duration of sunshine (hours)

N = Maximum possible duration of sunshine (hours)

a and b = constants (generally taken a = 0.25, b = 0.50)

The value of a and b is changes with place and values of a and b for some important place of India are given in Table 10.

Table 10 : The value of a and b at different station in India

Station	a	b	Station	a	b
New Delhi	0.256	0.454	Nagpur	0.227	0.459
Mumbai	0.283	0.397	Ahmadabad	0.252	0.448
Kolkatta	0.271	0.378	Jodhpur	0.313	0.401
Chenai	0.242	0.436	Goa	0.270	0.423
Pune	0.269	0.417	Shillong	0.225	0.463

The value of maximum possible sunshine hours at different latitude and important places of India is given in annexure- X & XII.

12.3.1. Pyranometer or solarimeter

Pyranometer is the instrument used to measure the solar irradiance incident on a horizontal surface from the entire hemisphere (Fig. 39). It is used for measuring global, diffused and reflected sky radiation reaching the earth's surface. When the instrument is directly exposed, gives global radiation; under inverted position, gives reflected radiation (short wave). To measure diffused radiations alone, solar direct rays are screened by a semi-circular shading ring and this instrument is know as solarimeter.

Fig. 39 : Pyranometer for measuring of direct solar and scattered sky radiation.

Multi-junction thermopile is generally used as a receiving surface. The sensor is coated with optical black and is mounted in a chromium plated brass case having concentric pair of hemispherical glass domes. When exposed, the blackned surface rises in temperature and produces thermal e.m.f. which is measured on a recording milli-volt meter or fed to recorder.

This instrument measures total short wave radiation or components on a horizontal plane surface from a hemispherical sky or 2 pi solid angle.

12.3.2. Pyranometer with shading ring

Pynarometer when shielded from direct solar beam by a circular band, measures only diffused radiation. The direct solar rays are screened by a semicircular shading ring mounted on the polar axis.

An occulating disc, tracking with the sun's motion, will cast shadow to cover glass hemispheres cutting direct beam and instrument will measure only diffuse component. Normally a shading ring is preferred than the occulating disc. For continuous record; arrangements of the rotation of the arm of the instrument is synchronished with the sun's apparent motion. The pyranometer is also used for the measurement of global radiation.

12.3.3. Pyrgeometer

The instrument measures the long wave outgoing terrestrial radiations from the earth surface (Fig. 40). The instrument consists of pairs of blackened and polished strips of mangnin. On expose to space, the dark strip cools more rapidly than the polished strip and by heating the dark strip electrically to the temperature of bright strip, a measure can obtain of a net out-going radiation.

This difference in temperature is sensed by the copper constantan thermocouple attached at the back of each of the strips and the thermo e.m.f. generated causes a deflection on a null galvanometer. The strips which are at lower temperature are heated by passing current till the galvanometer shows zero deflection on the scale. This indicates that the temperature of both the strips is same. The current required to compensate the heat loss is proportional to the net terrestrial radiant radiation.

12.3.3.1. Working principle

In the instrument blackened surface (inside a well polished case) is exposed to the radiation. The difference in temperature between the sensitive surface and reference points inside instruments case is measured by several thermo-junctions arranged in series. The blackened surface rises in temperature, until its rate of loss of heat by all causes is equal to the rate of heat gain by radiation. The rise in temperature sets up a thermal e.m.f. which is fed to a continuously recording mili-voltmeter (Fig. 41).

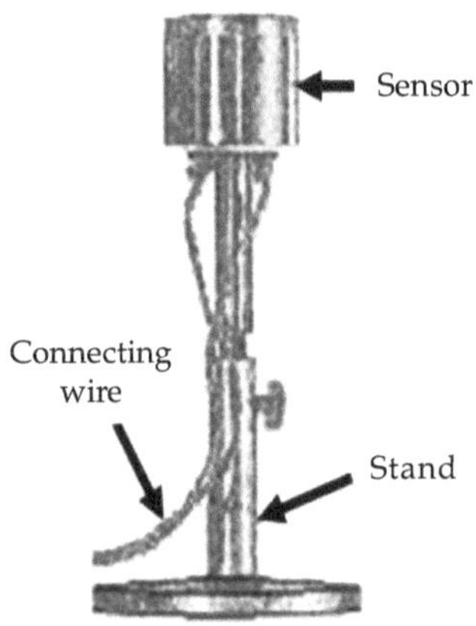

Fig. 40 : Pyrgeometer

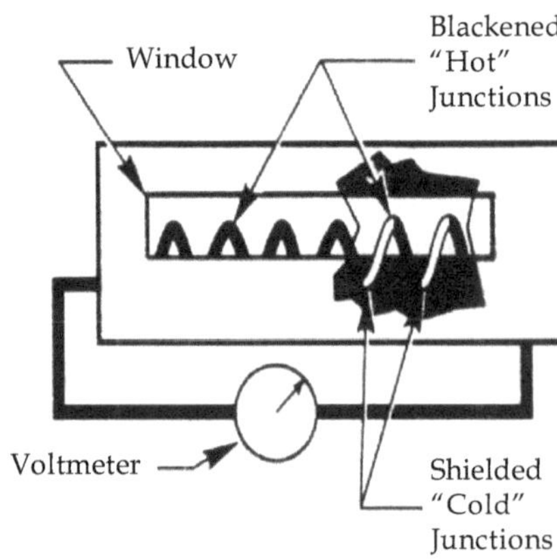

Fig. 41 : Thermopile of pyrgeometer

12.4. Precautions

- Pyranometers should be kept horizontal while in use.
- Observations should be repeated to check the accuracy.
- Any shading on sensors should be avoided.
- The window must be cleaned and inspected regularly, ideally every morning.
- A periodic check is to ensure that the silica gel desiccant is still coloured orange. When the yellow silica gel in the drying cartridge is turned completely transparent (normally after several months), it must be replaced by fresh silica gel.

1

It is another type of pyranometer, which is used to receive the reflected short wave radiation from earth's surface. Albedometer measures incident solar radiation as well as reflection radiation (short-wave) over different surfaces. The albedometer measures the net global radiation, as well as the albedo of grounds surface.

Fig. 42 : Albedometer

Two thermopiles are especially aligned to the same sensitivity, thus guaranteeing exact results. The thermopile of the instrument need a proper thermal insulation from the wind and to reduce the sensitivity to thermal irradiance, the albedometer are provided with a 4 mm thick glass dome having a 32 mm outer diameter.

The albedo meter consists of two hemispherical glass domes, one facing upwards and the other downwards, each containing the sensor (Fig. 42). In this instrument; 100 copper constantan thermocouples are used, embedded on a substrate containing AI_2O_3 which has higher thermal conductivity. The thermocouples are arranged circularly resulting in low azimuth error. Temperature coefficient of the receiver is thermistor-compensated, so that the sensitivity remains constant. Differential heating of the thermocouple junctions generates e.m.f., which is proportional to the incident energy and is measured as $mV\ s^{-1}m^{-1}$.

The sensitivity of the instrument is $4.71\ mV/Vm^{-2}$ and wave length range is 305-2800 nm. The albedo meter has the same sensitivity as that of the upper sensor. The following code is followed for identifying the terminals:

Upper sensor	-	green +	yellow -
Lower sensor	-	white +	brown -

The lower screen prevents illumination of the lower inner glass dome directly by the sun, near sunrise and sunset. When the outputs of the two sensors are recorded separately, the albedo of the underlying surface can be calculated. With the sensors connected in series opposite to each other, the short wave balance (net global radiation) can be obtained.

12.5.1. Exposure

The instrument is usually exposed at a height of 50-100 cm (generally 1 meter) above the surface over which measurements are to be made. No physical obstruction should be present within three times the height of exposure. The output can be measured by a micro or milli-voltmeter, which can be later converted to W/m^2 using the conversion factor provided for the instrument. Observations are to be taken preferably at hourly intervals from sunrise to sunset, to study diurnal variation of incident solar radiation and albedo, on horizontal surface. Differences in albedo over various natural surfaces can be studied and compared. Similarly the direct and diffuse components of global radiation and their diurnal variations, as well as the profiles of radiation in crop canopies can be studied with the instrument.

12.5.2. Observation

Keep the instrument on plain ground with the help of retort stand and see that it is levelled. Avoid obstruction from the sun side. Take the reading of total solar radiation, when the sun is shining bright without clouds. Note the reading in mV with the multimeter provided. Then shade the instrument from above and take the reading again, to get the diffuse radiation.

To measure the albedo from the ground surface, connect white and brown terminals to the multimeter and note the reading. Convert these reading into radiation units by multiplying with the constant of the instrument.

After taking the readings over bare ground, shift the instrument to crop field or to lawn and repeat the process again to get the albedo of the underlying crop canopy or grass surface.

To study the diurnal variation of both solar radiation and albedo, take the readings for every hour. Finally draw graphs to show the diurnal variation of solar radiation and albedo.

12.5.3. Working principle

The albedometer is based on a pair of thermopile pyranometers, one of them measuring the incident radiation on the ground [↓] and the other one the reflected radiation [↑]. The thermopile sensitive surface is coated with a black matt paint, which allows the pyranometer not to be selective at different wave lengths (Fig. 43).

The pyranometer spectral range is determined by the transmittance of the two glass domes type. Radiant energy is absorbed by the thermopile black surface, creating a difference of temperature between the center of the thermopile (hot junction) and the pyranometer body (cold junction). The difference of temperature between hot and cold junction is converted into a Potential Difference.

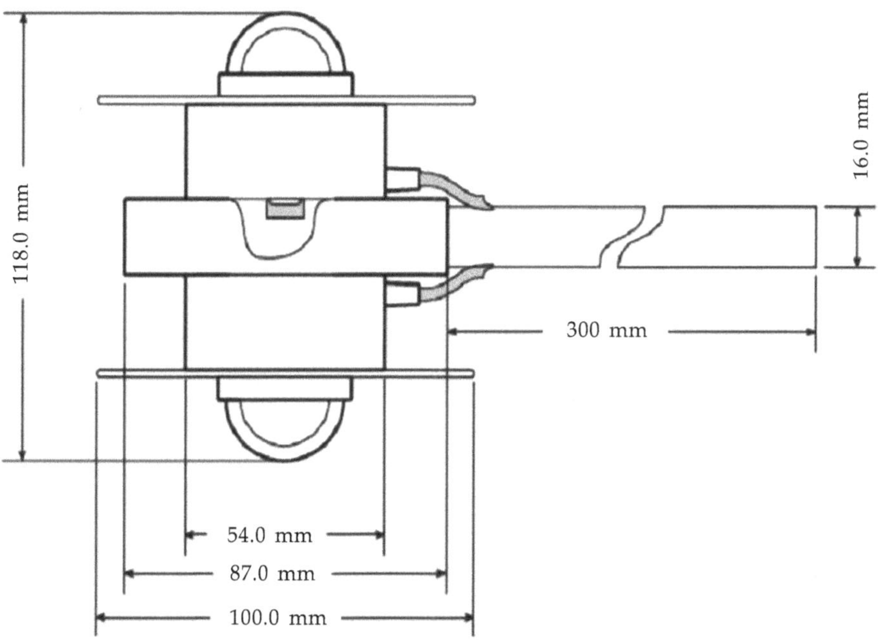

Fig. 43 : Inner view of Albedometer

12.5.4. Installation of the albedometer

The albedometer has to be mounted in a readily accessible location and it is generally mounted at 1-2 m above the ground. Grass should be kept at the same height all the year long. The albedometer has to be positioned in such a way that the power lines of the two pyranometers be pointed to the North Pole, if the instrument is used in the Northern

Hemisphere, and to the Southern pole, if used in the Southern Hemisphere. The pyranometer is provided with a spirit level for carrying out an accurate horizontal leveling.

12.5.5. Precaution

- The thermopiles of the instrument are protected against wind and weather.
- The domes protect the thermopile from the dust, which, laying down on the black surface, might change the spectral sensitivity. It is important to keep the outer glass domes clean to grant measurement with best accuracy.
- Washing can be made using water and standard papers for lens, or, in some cases, using pure ETHYL alcohol. After using alcohol, clean again the dome with water only.

12.6. Pyrheliometers

A pyrheliometer is used to measure direct component of solar irradiation at normal incidence (Fig. 44). The instrument is basically thermoelectric radiation receiver. Thermoelectric pyrheliometers use thermopiles, mostly copper-constantan.

Thermopile is mounted at one end of a long metallic tube provided with a series view limiting diaphrams. The front aperture is usually sealed with optical glass or quartz. The instrument is directed towards sun and a beam at normal incidence on the thermopile is obtained. After some time a constant deflection is obtained on a micro-voltmeter attached to thermopile. This reading in milli-volt is converted to (mw/cm^2) by multiplying a conversion factor. The viewing diphram limits the aperture to 5°. The instrument is sensitive to the total solar radiation in the spectral range 0.3 to 15 mm at intensities 10 to 200 mw/cm^2.

Fig. 44 : Pyrheliometer for the measurement of direct solar radiation at normal incidence.

12.6.1. Working principle

The detector of the pyrheliometer is based on a passive thermal sensing element called a thermopile. The thermopile responds to the total power absorbed by the black surface coating, which is a non-spectrally selective paint, and warm up. The heat generated flows through a thermal resistance to the heat-sink (the pyrheliometer body). The temperature difference across the thermal resistance of the detector is converted into a voltage as a linear function of the absorbed solar irradiance.

12.7. Net Pyrradiometer

Net radiometers measure the balance between the total incoming and outgoing radiation fluxes at all wavelengths (Fig. 45). The principal component of net radiometers is a flat black thermopile plate exposed to radiant energy measures the downward, upward and net total radiant energy. The thermopiles are covered using polythene domes having transparency in the range 3 to 80 mm with some narrow absorption in between. To overcome the effect of the wind, the plate is ventilated at a constant rate with dry air or nitrogen gas.

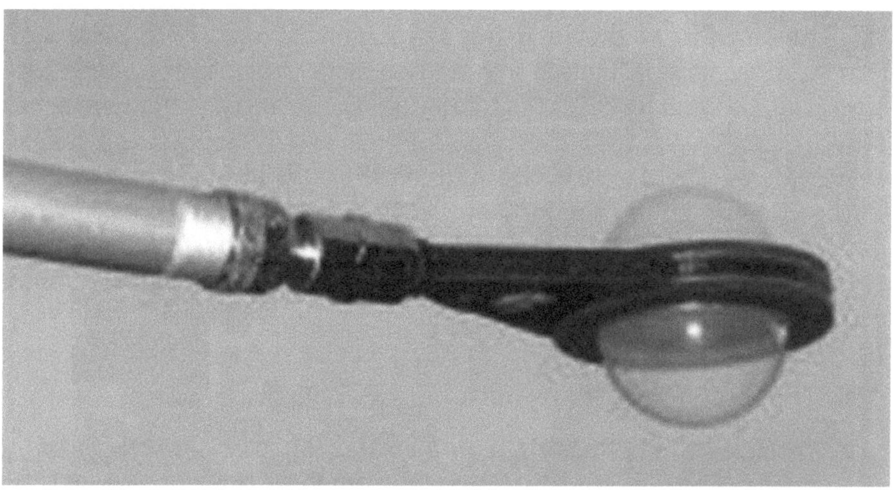

Fig. 45 : Net Pyrradiometer

All net radiometers require a milli-voltmeter for output measurement and a power supply to operate the ventilating fan or pump for regular circulation of dry gas or air to protect the instrument from any excess heat development.

12.7.1. Working principle

Net radiometer is based on a thermopile sensor with one set of hot junctions in contact with the upper surface and a set of cold junctions in contact with the lower surface. The difference in temperature between the two receivers is proportional to the net radiation. Through the Seebeck effect, the difference in temperature between hot and cold junctions is translated into a Potential Difference.

The thermopile sensing element is made up of a large number of thermocouple junction pairs connected electrically in series. The absorption of thermal radiation by one of the thermocouple junctions, called the active (or 'hot') junction, increases its temperature. The differential temperature between the active junction and a reference ('cold') junction kept at a fixed temperature produces an electromotive force directly proportional to the differential temperature created. The instrument is provided with window shields (transparent) with forced ventilation which prevents wind effect, deposition of dust, and deposition water over the sensor. Dry air is circulated at a slow rate, just keep the hemispherical domes fully inflated.

12.7.2. Observation

The sensors of the instrument are sensitive to both solar and terrestrial radiations. Unlike pyranometers, the net pyrradiometers are uniformly painted black on the two sides - the surface facing up receives downward energy and the one facing down receives the upward energy. By proper electrical connections, the outputs of the two surfaces can be measured individually as well as the differential outputs between the two surfaces. Take the reading of the difference in temperature between the two receivers, which is proportional to the net radiation.

12.7.3. Precautions

- The receiving surfaces must be always kept clean.
- Washing can be made with water and standard lens paper or by pure ETHIL alcohol. After using alcohol, the domes must be washed with water only.
- Yearly calibration of instrument is required.
- Calibration can be checked directly in the field, by mounting another net radiometer (sample) alongside, and comparing results.

Questions for Study

Q.1. What is difference between radiation and albedo?

Q.2. Write the working principle of Pyranometer and net radiometer?

Q.3. What are the precautions to be adopted for Pyranometer and net radiometer?

Q.4. What type of exposure is required for albedometer?

Q.5. How global radiation is estimated from sunshine hour values?

❑❑❑

Chapter 13

Measurement of Evaporation

The change of state of water form solid or liquid to gaseous form is known as evaporation. It may be defined as the water loss in vapour form from a soil/water surface. The energy used in this process is called as latent heat. The energy of 540 cal/g of water is used in this process. This process plays a major role in the redistribution of thermal energy between the earth and the atmosphere. Water vapour diffuses into the atmosphere from the different natural surfaces like ocean, river, lake, pond, soil and plant etc. It is difficult to measure the evaporation from the natural surfaces. To measure the evaporation from open water surface is easy and is normally in practice. The rate of evaporation is dependent on the size of the measuring surface (Pan) at constant wind speed below saturation point. Therefore a measured surface (Pan) is designed to measure the evaporation.

13.1. Measuring Instrument

USWB-Class A open Pan Evaporimeter is most commonly used in observatories for measuring evaporation (Fig. 46).

13.1.1. USWB-class A open pan evaporimeter

Evaporimeter is an instrument used to measure evaporation over an area in a specified time. It measures the evaporating power of air layers near the ground. Evaporation is usually expressed in units of depth of water lost in mm per day. The open pan evaporimeter do not provide directly the evaporation from a natural surface. It indicates only, the order of magnitude of evaporation from such surfaces.

13.1.1.1. Working principle

The amount of water lost by evaporation from pan in any given interval of time is measured by adding known quantity of water to the pan from a graduated cylinder till the water level touches the reference point.

Fig. 46 : USWB - Class A open Pan evaporimeter

13.1.1.2. Description of USWB - class A open pan evaporimeter

The different parts of USWB-Class A open Pan Evaporimeter (Fig. 47), most commonly used in observatories for measuring evaporation by IMD are described below.

Pan

A big circular pan is required to store water for free evaporation. The pan is made of 22 gauge galvanized iron or copper sheet. The diameter of pan is 122 cm and height of 25 cm.

Still well

Winds disturber the water level and create ripple on the water surface in the pan. To get undisturbed water level and to break any ripple generated by wind, a still well is placed inside evaporimeter. Still well is made up of a brass cylinder mounted on a heavy circular base (Fig. 48). It consists of a brass rod which is fixed to the center of the still well with sharp tip, which provides the reference point up to which water level is kept at the time of observation. The tip of the brass rod is about 19 cm above the base of the pan. To communicate the water from pan to the still well there are 3 small holes (120 degrees apart each) at the bottom of the still well.

Mesh Cover

The pan is covered with iron mesh to avoid that loss of water from the pan due to animals and birds.

Thermometer

A thermometer is placed in the pan to measure the water temperature at the time of observation. It is fixed by a clamp to the side of the pan and bulb of thermometer is 5 cm below water surface.

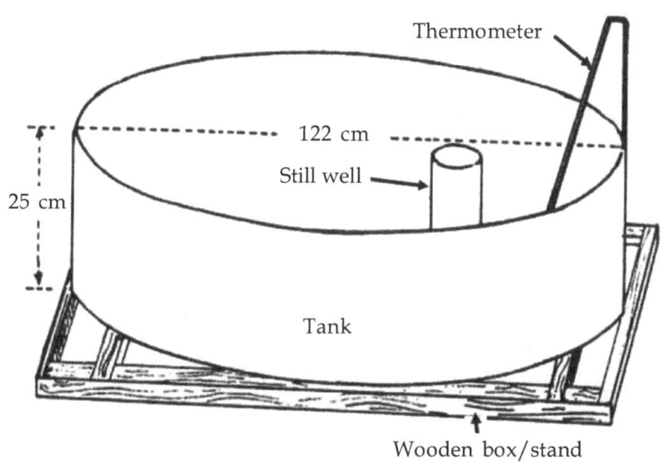

Fig. 47 : Different parts of USWB-Class A open Pan Evaporimeter

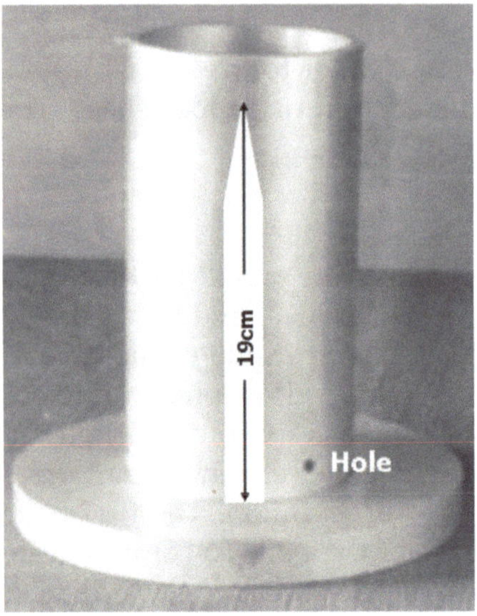

Fig. 48 : Still well

Fig. 49 : Measuring cylinder

Measuring cylinder

The measuring cylinder is made of brass and had graduated brazed scale from 0 to 20 cm from top to bottom (Fig. 49). The scale is fixed on the inner side of the cylinder. There mark in this scale at every 1 cm and it has accuracy of the measurement of 0.1mm. It has a diameter of $1/10^{th}$ that of pan so that the cross sectional area of the cylinder is exactly $1/100^{th}$ that of the pan.

13.1.1.3. Time and method of observation

The observation is recorded daily at 8:30 AM and 2.30 PM. Water is added with measuring cylinder to the pan until the tip of still well coincides with the surface of water in stilling well. Total evaporation is calculated by subtracting today's reading from the previous day readings. If rain has fallen since the time of the previous observation the level of the water in pan rises. The amount of evaporation is measured with a graduated cylinder i.e., the amount of added water to get accurate level of water with the fixed point gauge is evaporation.

Thus 200 mm of water added by one full cylinder will raise water level by 2 mm in pan.

$$\text{Therefore, Evaporation} = \frac{\text{Total volume of water added by cylinder mm}}{100}$$

If rainfall has occurred after previous reading and exceeds the water lost by evaporation then water has to be removed from the pan with the help of same cylinder instead of adding water.

13.1.1.4. Installation of USWB - class A open pan evaporimeter

A platform of broken bricks, stones and bricks is prepared according to the measurements given in figure 50. After putting a layer of 2 inches of loose soil on the platform, the surface is made hard and the evaporimeter is placed upon a wooden stand of 4' x 4' x 4" and this ensures free circulation of air around the pan. The wooden stand avoids the direct contact of the pan to the soil (remove the soil temperature effect) and also does not get tilt after rains.

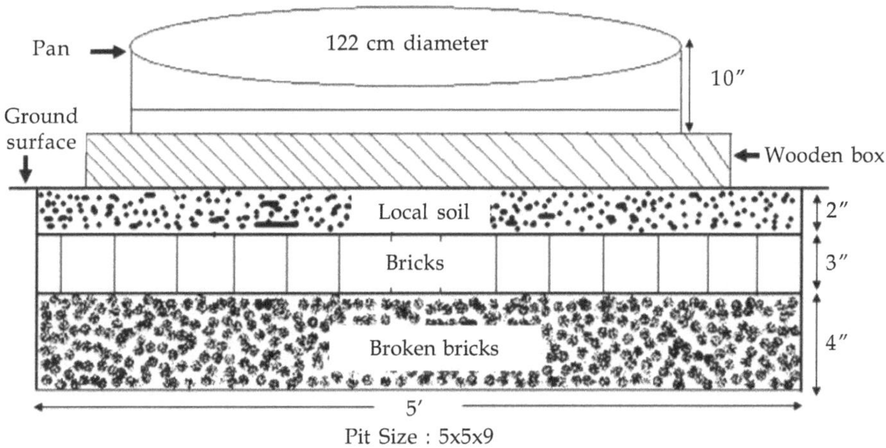

Fig. 50 : Installation of USWB - Class A open pan evaporimeter

The pan should be placed on the wooden board and bottom of the pan is perfectly horizontal. The level of the top rim of stilling well should also be checked with a spirit level. The stilling well is placed in the evaporimeter at 30 cm away from the North edge of the pan to facilitate viewing of the water level. After this, a wire mesh should be placed over the pan tightly fitting the rim all around..

13.2. Precautions

- Pan should be placed on a horizontal wooden platform.
- Area around wooden platform should be kept weed free.
- The instrument should not be placed on a concrete base as it will heat up the pan and affecting the evaporation rate.

- Wooden planks should be painted white to reflect radiations falling on the instrument.
- Wire mesh should be tight and white painted.
- Temperature of evaporating surface should be recorded.
- It should be located at a site which is leveled and far from any obstructions.

13.3. Measurement of Daily Evaporation

The following examples would illustrate these procedures:

To measure evaporation, a measured volume of water is added with the help of measuring cylinder so that the original level of water to the tip of reference point is restored.

I. Water added at 0830 hrs IST on 4-11-1982 to bring the water level to the reference point is 51 cm. Rainfall from 0830 hrs IST on 3-11-82 to 0830 hrs IST on 4-11-82 is Nil. Evaporation on 4-11-1982 = 51 cm/100 = 5.1 mm.

II. Water removed at 0830 hrs IST of 10-7-82 = 64 cm. This amounts to a change in water level of 6.4 mm. The rainfall recorded at 0830 hrs IST of 10-7-1982 is 8.4 mm. Had there been no rain, the level would have risen by 8.4 mm. So the evaporation for the day is 8.4- 6.4 = 2.0 mm.

III. The water added at 0830 hrs IST on 5-6-82 is 23 cm. This amounts to a change in water level of 2.3 mm. If the rainfall recorded at 0830 hrs IST on 5-6-1982 is 1.6 mm, the actual evaporation is equivalent to the sum of rainfall and the decrease in water level. So the evaporation for the day: 2.3 + 1.6 = 3.9 mm.

13.4. Solved Exercise of Evaporation

Normally four situations occur while recording evaporation from the Class A open pan evaporimeter.

Situation 1 : Day with no rainfall

Under this situation, a known quantity of water is added to the evaporimeter to bring the water level to the reference point.

Exercise : Calculate the evaporation if two full cylinder of water added into the evaporimeter in non rainy day to bring the level to reference point.

Solution : Water added to the pan = 2 full cylinder
One full cylinder of water = 2 mm of water evaporated

Therefore, Evaporation = 2 x 2 mm
= 4.0 mm for that particular day

Situation 2 : Day with light rainfall

If there is any light rainfall, the evaporimeter will also receive the rain water, since it is open. Under this situation, water added through rainfall should be accounted while recording evaporation.

Exercise : Calculate the evaporation if water added to the evaporimeter to bring the level to reference point is equal to 2 full cylinders and the rainfall for that particular day is 2.0mm.

Solution : Rainfall for that particular day = 2.0 mm
Water added to the pan = 2 full cylinder
One full cylinder of water = 2 mm of water evaporated

Therefore, Evaporation = (2 x 2) + 2.0 mm
= 6.0 mm for that particular day

Situation 3 : Day with heavy rainfall

If there is heavy rainfall, the water level will go above the reference point in the evaporimeter. Under this situation, water should be removed from the evaporimeter to bring down the water level to the reference point.

Exercise : Calculate the evaporation if water removed from the evaporimeter to bring down the water level to reference point is equal to 2 cans and the rainfall for that particular day is 10.2 mm.

Solution : Rainfall for that particular day = 10.2 mm
Water removed from the pan = 2 full cylinder
One full cylinder of water = 2 mm of water evaporated

Therefore, Evaporation = 10.2 - (2 x 2) mm
= 10.2 – 4 mm
= 6.2 mm for that particular day

Situation 4 : Day with very heavy rainfall

If there is very heavy rainfall, the water level goes beyond the capacity of the pan and overflow occurs. Under this situation, no known standard method is available to measure the evaporation and it is recorded as no evaporation or 'flooded'.

Questions for Study

Q.1. What is evaporation and the factors that influence it?

Q.2. Discuss the precautions and procedure for measurement of evaporation.

Q.3. Why open pan evaporimeter is painted white from outside and tinned inside every year?

Q.4. What is least count value of the hook gauge?

Q.5. On a rainy day if tank is filled and over flow with water, how much evaporation is expected?

❑❑❑

Chapter 14

Measurement of Dew

Dew acts as a source of moisture for the crop plants. It causes reduction in transpiration rate and saving of soil moisture reserves. Dew fall helps in insect-pest and diseases spread in crop plants.

14.1 Dew and Conditions for Dewfall

The radiational cooling during night causes the moisture to saturate and further cooling results into condensation and depositing this moisture in the form of tiny droplets on the grasses, vegetative and other non-conducting surfaces. This deposition form is known as dew. Clear nights, low wind velocity in the order of 1.0 to 3.0 m/s, presence of moderate to high moisture (RH = 75%) and sufficient radiational cooling are the four necessary and conductive conditions for the occurrence of dew.

14.2. Difference between Dew and Water of Guttation

Dew and water of guttation may appear similar on the deposition surface and hence create confuse. They are distinguished by careful observation and with help of photographs. Water of guttation is deposited on the tip of the leaves through physiological process, while the dew is deposited over the leaves through atmospheric process.

Dew droplets size is ranged between 0.02 and 0.35 mm. Guttation of water droplets assumes the size of 2 mm to 3 mm diameter, while dew particles seldom grow to that size and spread over the leaves. The size of guttation of water droplets is bigger in size as compared to the dew.

14.3. Dewfall

Dew deposition is a nocturnal phenomenon. When downward transported atmospheric moisture gets condensed on the cooled surface, it is called dewfall.

14.4. Measuring Instrument

The Duvdevani Dew Gauge is used to estimate dew manually at various heights from the ground (Fig. 51) and Dew Recorder is an automatic instrument which records dew duration and enables to find out amount and intensity of dew.

14.4.1. Description of duvdevani dew gauge

The dew gauge is wooden blocks, which was designed by an Israel scientist, Duvdevani in 1953. It consists of a rectangular wooden block of size 32 x 5.0 x 2.5 cm (Legth x width x height). The wooden blocks are coated with red oxide for the retention of dew deposited on it. The dew gauge is exposed on special iron supports on a wooden stand. The stand is fixed (Fig. 51) vertically on the ground and the iron supporters are fixed at heights 5, 25, 50 and 100 cm above ground. These supporters should be well leveled.

The support and clamps for the plate and gauge shall be of non-rusting, non-magnetic and non-corrosive stainless steel or similar material.

The supports shall be firmly levelled mounted on the vertical stand in opposite directions alternately by brass screws exactly at the heights of 5 cm, 25 cm, 50 cm and 100 cm from the ground level.

A groove each of 5mm x 10mm shall be made on side B at a distance of 48mm from either edges of the dew gauge throughout its width so that it site firmly on the support.

14.4.2. Working principle

Material, which differs in its thermal properties from the surrounding air and having smooth surface finished with good quality of red oxide paint, when exposed to very low temperature conditions in the open air, a dew is form on its smooth surface. The material shall preserve the dew fall some time after sunset so that observer can record it.

14.4.3. Installation of duvdevani dew gauge

Dew gauge plates are placed on an iron stand. The iron stand is fixed on a foundation pit size of 30 x 30 x 60cm (Fig. 52). After digging the pit size, the lower portion of 20cm portion filled with lime concrete. The dew gauge stand is fitted in the pit such a way that last clamp is fixed on 100cm from ground and then fill the pit with cement concrete upto 30cm in the ground level. The iron stand is in perpendicular position. Put 10cm loose soil layer around the iron stand.

Fig. 51 : Duvdevani Dew Gauge

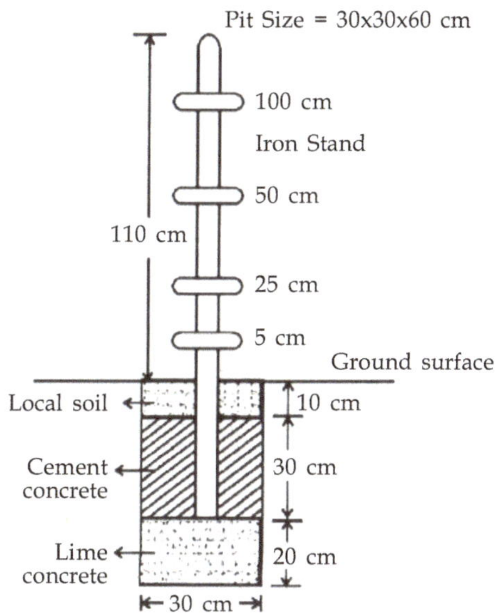

Fig. 52 : Installation of Duvdevani Dew Gauge

14.5. Method and Time of Observations

Everyday, just before sunset, the four gauges are placed on the exposure stand. The dew gauge with the highest number is placed at the highest level and subsequently other gauges in the descending order. The dew gauge is exposed at about sunset to at about sunrise. The reading of dew gauge is recorded before sunrise every day in winter season.

The size, form and distribution of the dew deposited on the gauge are observed carefully. This appearance of dew is compared with a set of photographs of each dew type supplied in photo-album (Fig. 53a & 53b). Examine the dew formed and select the photograph which most nearly corresponds to its appearance. Note down the correct dew scale. If the drop patches are not similar to any of the photograph, enter 'X'. The water equivalent to the dew scale is noted in observation Table (Table 12).

14.6. Dew Album

It is an album containing photographs with numbers 1 to 10, out of which 1 to 8 refer to dew, 9 refers to dew and rain together; and 10 refers to rain (Fig.53a & 53b). The set of photograph bear its equivalent quantity of dew in 'mm'.

14.7. Estimation of Dewfall

The size, form and distribution of dew drops on the surface of the plates are compared with a series of standard full scale photographs and dew scale number is given for the amount of dew present on the surface of the wooden plates in the morning; first of all we should record the reading of dew gauge before sunrise. The dew deposit on the dew gauge surface is examined and the corresponding dew scale is noted.

The water equivalents in mm of these dew scales vary from 0.02 mm for dew scale number 1 to 0.35 mm for dew scale number 8 (Table 11). Thus, knowing the dew scale number, it is possible to measure dew. The drop of rain is of spreading nature whereas; the drop of dew is some thing like pearl shape.

Table 11 : Dew scale number with dew equivalent in mm

Dew Scale		Remarks
Dew scale number	**Dew equivalent in mm**	
0	No dew	More than one photographs (a, b) are given for some numbers to represent dew deposits that fall within the same dew scale number e.g. 2a, 2b etc.
1	0.02	
2	0.045	
3	0.075	
4	0.11	
5	0.16	
6	0.21	
7	0.27	
8	0.35	
9	No observation	

DEW ALBUM

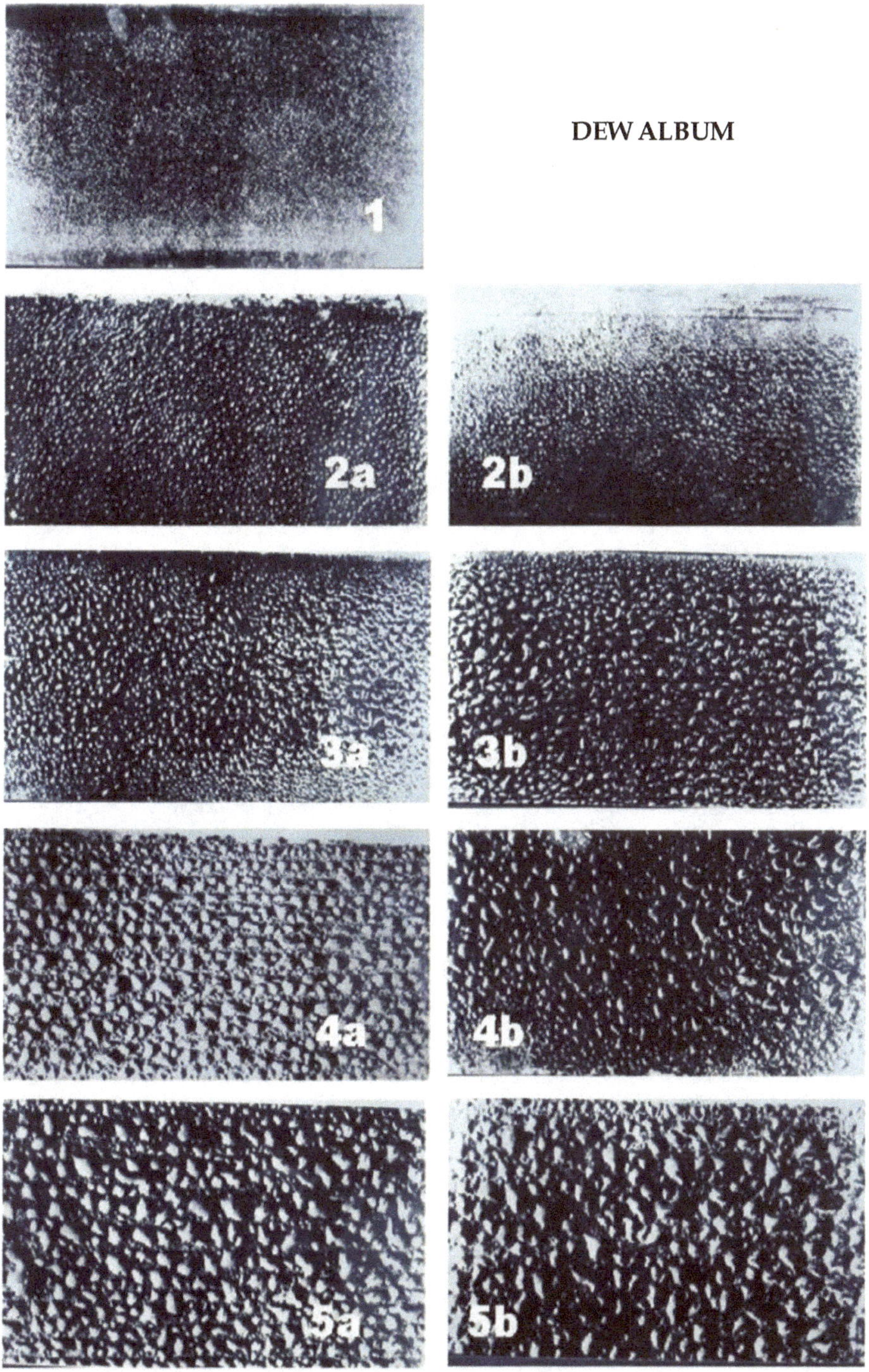

Fig. 53a : Dew particle photograph with dew scale

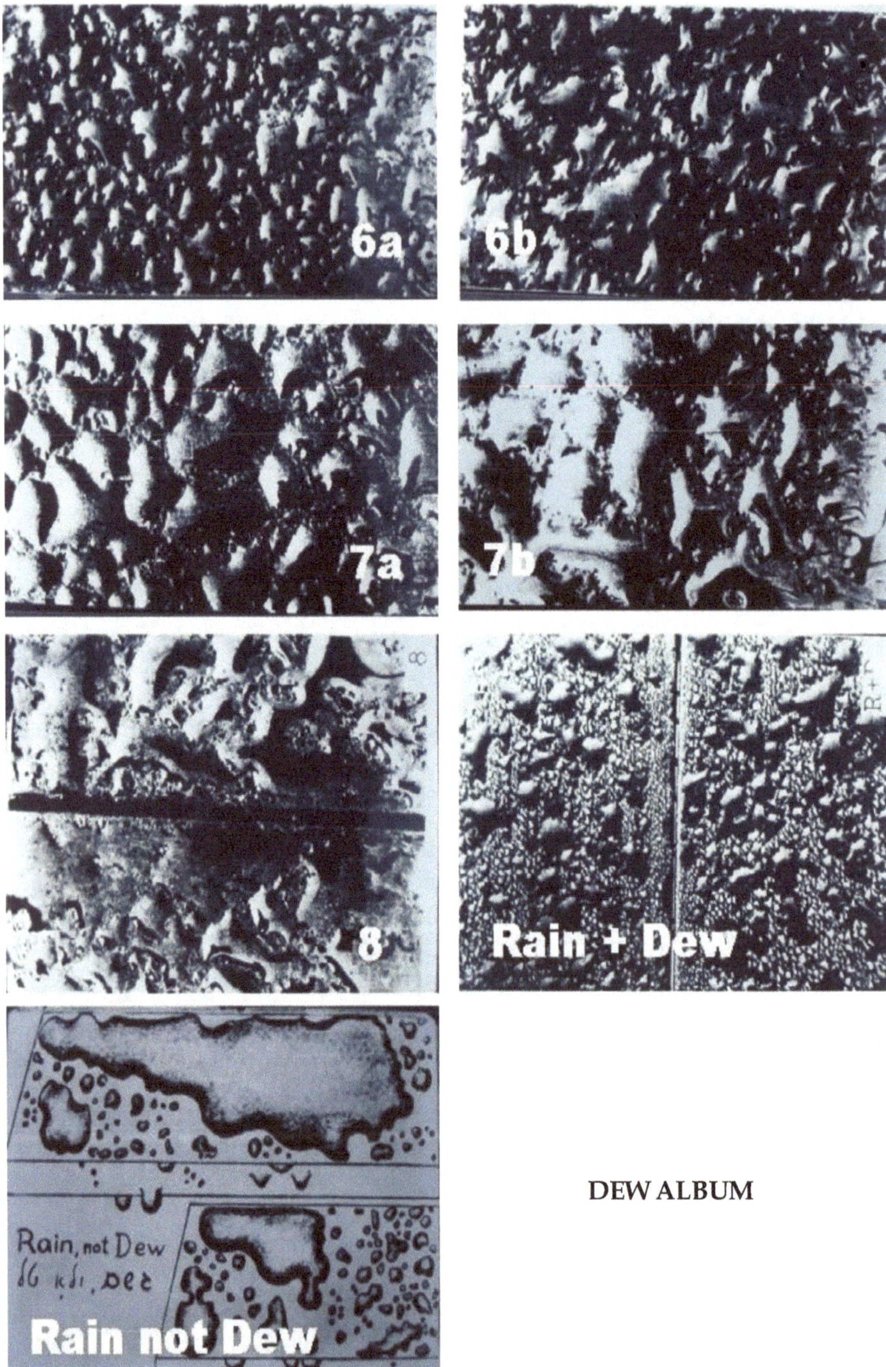

DEW ALBUM

Fig. 53b : Dew particle photograph with dew scale

Table 12 : Observation Table

Observation time : Just before sunrise

Date and month	Amount of dew (mm) average of top and bottom at the height of				No. of dew nights (monthly)			
	5 cm	25 cm	50 cm	100 cm	5 cm	25 cm	50cm	100 cm

14.8. Precautions

- Do not touch the surface of dew gauge with grease of dusty fingers.
- The surface should not be rubbed. The surface of the dew gauge should be cleaned in running water. It is vigorously shaken and allowed to dry.
- After recording the dew observation, keep the dew gauge indoors otherwise sunlight will damage the sensitive surface of the dew gauge.
- After taking the observation, do not wipe off the water deposited but remove by swinging the gauge.
- The sunlight may interfere with photograph comparison, therefore dew must be observed under the observer's body shadow.
- Additional information like rain, frost, fog etc. should also be recorded.
- It should be applied fresh coat of red oxide as per needs.

Questions for Study

Q.1. How dew is formed and what are the meteorological conditions associated with the formation of dew?

Q.2. What is frost and how it forms?

Q.3. How is dew measured?

Q.4. Discuss the precautions and procedure for estimating dew amount.

❑❑❑

Chapter 15

Measurement of Pressure

Pressure of air at a given place is the force exerted against a surface by continuous collision of gas molecules. Pressure may be defined as the force per unit surface area.

$$P = \frac{F}{A}$$

where, P = Pressure
F = Force
A = Area upon which force is exerted

Air pressure is the force exerted in all directions as a result of weight of all the air above it. Force exerted by mass of a column of air above a given point is called air pressure.

$$P = \frac{\text{Mass} \times \text{Acceleration}}{\text{Area}} \quad [F = m \times a]$$

where, Mass = $\rho \times V = \rho \times h \times A$

$$P = \frac{\rho\, h\, A.g}{A} = \rho h g$$

where,

ρ = Density of liquid (Mercury = 1.3 ×103 kg m^{-3})

h = Height of liquid (Mercury height = 76 cm)

g = Acceleration due to gravity (980 cm s^{-2})

P = 13.6 ×103 kg m^{-3} × 0.76 m × 9.8 m s^{-2}

= 101300 (kg m s^{-2}) m^{-2} [1N = 1 kg ms^{-2}]

= 101300 N m^{-2}

= 101300 Pa = 1.013 ×10^5 Pa

15.1. Units of Pressure

1. *Height of Hg* i.e. inches, cm or mm of Hg.
2. *Bar:* It is the force equal to 106 dynes per square centimeter.
3. *Pascle (Pa):* It is the force of 1 Newton per square meter = N/m^2
4. Atmosphere = 1 atm = 29.92 inches or 76 cm or 760 mm of Hg

   = 1013 milli bar (mb)

   = 101325 pascle = 101.3 kilo pascle (KPa)

   = 14.7 lb/$inch^2$

   = 1.013 × 106 dynes/cm^2

   = 1.013 bar

   = 1034 gms/cm^2

   = 1.013 x 105 N m^{-2}

   1" = 33.86 mb

   One pressure unit can be converted into other pressure units.

15.2. Standard Atmospheric Pressure

It is the pressure exerted by the atmosphere measured at temperature 273°K at mean sea level and at 45° latitude. Its value is 29.925 inches or 760 mm or its equivalent; 1013.25 mb. This is also called 1 atmosphere.

15.3. Measuring Instrument

Instrument used for measuring atmospheric pressure is called barometer. Following are the different types of barometers:

1. Fortin's barometer
2. Aneroid barometer
3. Barograph

15.3.1. Working principle of mercury barometer

When an air evacuated long and hallow glass tube is immerged with its open end down in mercury filled cistern, than the pressure of the outside air forces the mercury in the cistern upward into vacuum chamber within the tube. Under standard conditions of sea level; the atmosphere balances a column of mercury 76 cm height. As atmospheric pressure changes, the height of the mercury will fluctuate accordingly (Fig. 54). Determination of actual pressure requires corrections for temperature, instrumental error, and variation of gravitational force with latitude and altitude. The corrected pressure is the actual pressure of a place.

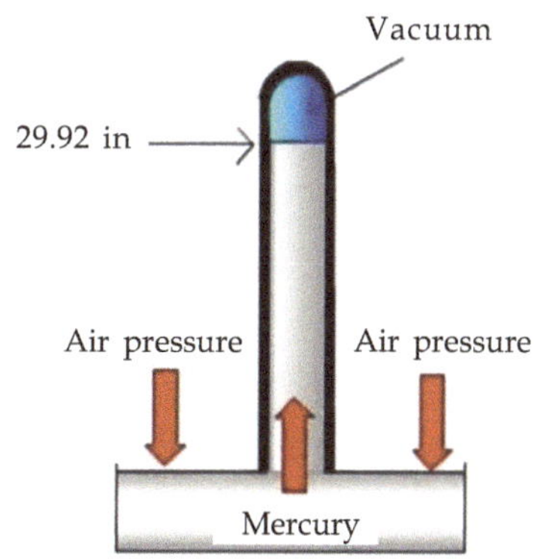

Fig. 54 : Mercury barometer

15.4. Description of Different Barometers

15.4.1. Fortin's barometer

This type of barometer is not easily portable but is capable of recording accurate atmospheric pressure. It has a closed glass tube with its lower open end immersed in a cistern of mercury. An adjusting screw located at the bottom of the lather bag that contains mercury can change the mercury level in cistern (Fig 55). The mercury in cistern is adjusted to the zero point (the ivory point). The convex meniscus of the upper mercury surface in the barometer tube indicates the atmospheric pressure. Two air holes are provided to keep the mercury in the cistern under atmospheric pressure all the time. The top of cistern is covered by a boxwood bushing and sides are surrounded by a glass cylinder.

The upper portion of the barometer has a mercury thermometer, a scale and a vernier, all attached to a outer brass tube in which the barometric tube is enclosed. In the upper part of the brass tube, two long slots are cut opposite one another to permit viewing of the upper level of the mercury column. The flexible bottom of cistern is adjusted with the help of a screw to fix zero level above which the height of mercury column is measured. The screw is moved until the surface of mercury in cistern touches the end of a fixed ivory point. The level of mercury in tube may be visible clearly through a slit to which a vernier system is attached to ensure correct reading. The scales of air pressure are marked on side of the slit.

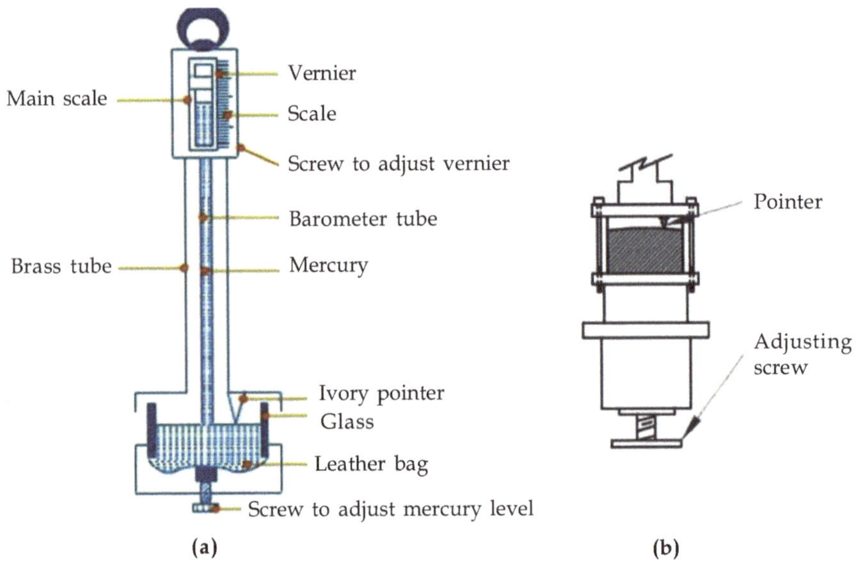

Fig. 55 (a & b): Fortin's barometer and their parts

15.4.1.1. Observation

To take readings from the barometer, keep it exactly in vertical position in cistern with the help of adjusting screw so as to touch the tip of ivory point. After this, read the main and vernier scales reading. After taking these two readings, accurate reading of pressure may be obtained by apply necessary corrections for temperature, gravity, altitude and latitude.

1) The variation in length of metal and mercury scale is changed with the change in temperature. The scale is engraved at a certain temperature and usually the readings are required to be taken at some other temperature. All readings are reduced to a standard temperature of 32°F or 0°C. Note down the temperature at the time of reading. There are correction Tables for temperature which can be used and for the same is done using following equation also.

$H_0 = H_t\{1-(\beta-\alpha)t\}$

where

H_0 = Corrected mercury height at 0°C

H_t = Observed mercury height at temperature t°C

α = 0.000019/°C, β = 0.00018/°C

t = Room temperature °C

2) Barometer height must be stated for a standard value of 'g' i.e., 980.656 cm/sec^2 because the acceleration due to gravity is not same at all the points on the earth's surface.

3) Effect of surface tension of mercury in a small bore tube needs a correction factor.

4) Effect of latitude should be considered. The observed value is reduced to a standard value taken at 45° N latitude.

15.4.1.2. Precautions

- Keep barometer exactly in vertical position.
- Adjust level of mercury in cistern with the help of adjusting screw so as to touch the tip of ivory point.
- Read main and vernier scales readings correct.
- Apply necessary correction.
- The convex meniscus is depressed due to surface tension. Hence, a tube of about 1 cm bore is taken when the effect of surface tension become negligible small.

15.4.2. Aneroid barometer

This barometer is a simple portable, convenient to use and without mercury barometer. It consists of sylphon chamber which is partially evacuated cylindrical box corrugated at its top and side. This chamber is made up of beryllium-copper, which is a sensitive element to atmospheric pressure variation. When the pressure increases, the box is compressed and when pressure decreases, the box expands. A strong spring within the chamber avoids it from collapse under air pressure. The changes in the sylphon chamber are very small and are magnified with the help of levers and are communicated through chain and pully to the pointer. The pointer moves around a dial that is calibrated in the pressure unit (Fig. 56). This pointer gives direct reading in mm or millibar of atmospheric pressure.

15.4.2.1. Working principle

Some of the mixtures of two dissimilar elements like beryllium-copper, german-silver are very flexible and very sensitive to atmospheric pressure. A vacuum chamber is made up with this sensitive elements and suitable mechanical device is attached to the chamber to record the change in air pressure. When the pressure increases, the box is compressed and when pressure decreases, the box expands and pressure is read directly.

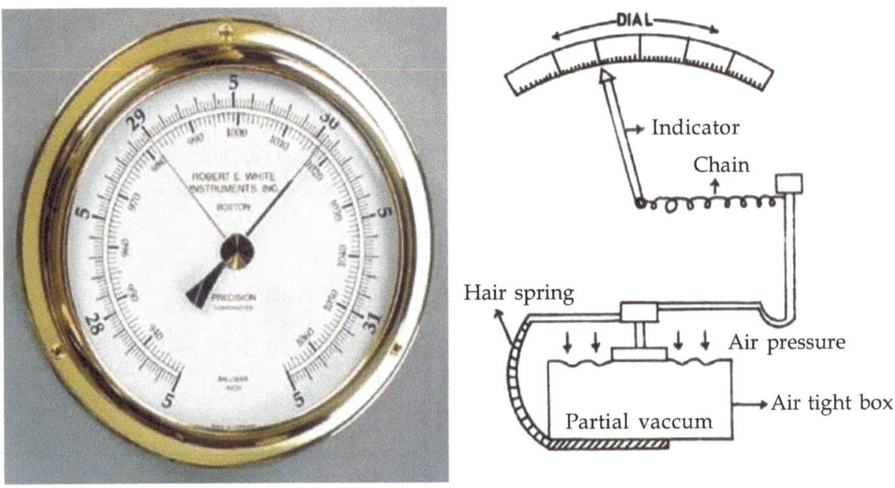

Fig. 56: Aneroid barometer and their parts

15.4.2.2. Precaution

The barometer must be set and standardized with the help of standard mercury barometer.

15.4.3. Barograph

This instrument is simply the integration of a number of aneroid barometers. It is an automatic instrument which is used for continuous record of atmospheric pressure with time. It consists of a battery of six to ten vaccum boxes of aneroid barometer - placed one above another. The number of vacuum boxes is used to reduce the effect of irregularities and to make the instrument more sensitive (Fig. 57). The motion of these boxes is communicated by a system of levers to an arm of a recording system which carries pen with ink. As the pressure changes, the pen moves up and down making marking on the graph paper which is wrapped around the rotating drum. The drum is rotated by a clock work mechanism. The pressure is recorded to the precision of one millibar. This instrument is not accurate and needs its standardization before its use. The graph is to be replaced everyday.

The reading of barometer is to be made at the exact hours concerned at routine 8:30 and 14:30 IST hours or at desired hours.

Fig. 57: Barograph for recording automatic pressure variation with time

15.4.3.1. Precaution

- Set the reading before use with standard barometer.
- Use proper graph and replaced it every day.
- Check the pen and ink on weekly basis.

Question for Study

Q.1. What is standard atmospheric pressure?

Q.2. Which barometer is more accurate in comparison to others?

Q.3. How barometer helps in weather forecasting?

Q.4. How barograph works?

Q.5. Write the working principle of aneroid barometer.

Q.6. How continuous record of pressure is recorded?

❑❑❑

Chapter 16

Measurement of Photosynthetically Active Radiation (PAR)

Photosynthetically active radiation (PAR) is the radiation in the 400-700 nm wave band. For plants, 400-700 nm wave band of the light spectrum is very useful because of photosynthesis activity in this band. A simple integral relationship exists between the number of molecules changed photo-chemically and the photons absorbed within the requisite wave bands regardless of photon energy. Net photosynthesis rates are found to be almost linearly proportional to the radiation intercepted, which can be measured by line quantum sensor.

16.1. Measuring Instrument

Quantum/Line quantum sensors measure PAR in terms of Photosynthetic Photon Flux Density (PPFD). It is the number of photons in the 400-700 nm wave band incidents per unit time on a unit surface. Quantum/Line quantum sensor responds equally for all photons in the 400-700 nm wave bands and has a cosine response with a sharp cutoff at 700 nm.

16.1.1. Description of line quantum sensor

The line quantum sensor is usually of one meter in length and 12.7 mm wide (Fig. 58). It is composed of an array of high stability silicone photo voltaic detectors placed 2.38 cm apart. These detectors are placed in a water proof anodized aluminum case with acrylic diffuser and stainless cover. It has a sensitivity of 7 mA per 1000 mole per second per m^2. It has a very fast response time of 10 micro seconds and a very small temperature sensitivity of 0.15% per °C. It does not have any effect of tilt over 45 to 360°.

Fig. 58: Line quantum sensor

The sensor output is connected to an integrating quantum meter along with a calibrated connector (provided with each sensor). The meter has a provision to read instantaneous and also values integrated over 10, 100, 1000 seconds to give average values over the set time periods. These average values remove the effects due to the rapidly changing radiation because of cloud cover, surface waves and movement of sensor. The readout is by liquid crystal display (LCD), which eliminates linearity and interpolating problems. It has a memory to hold a particular reading as long as possible. The instrument has an option of connecting to a continuous recorder. Quantum sensor output is given in $mE/s/m^2$. The readings can be converted to W/m^2 by multiplying the reading in $mE/s/m^2$ by 0.2174 for natural daylight conditions.

16.1.2. Observation

The line quantum sensor is placed perpendicular to the rows of the crop horizontally at the ground level to get the transmitted PAR (TPAR) and above canopy or in the open to get incident PAR (IPAR), the sensor is inverted over the crop canopy to get the reflected PAR (RPAR). The difference between IPAR and the sum of RPAR and TPAR give the absorbed PAR (APAR). Generally these values are expressed in terms of per cent of incident PAR. To measure the profile IPAR, the sensors are placed at 25%, 50% and 75% of the crop height. If one has to take one time reading, it is advisable to take the readings between 1100 to 1300 hours. But while taking diurnal observations, hourly readings are advisable from sunrise to sunset. Frequency of observations always depends upon the objectives of the experiment.

The line quantum sensor is placed perpendicular to the crop rows to get the TPAR. This depends mostly on the extent of leaf area or canopy cover. The reading of TPAR gives the extent of PAR utilized by the crop. The integrator is used to take the integrated readings on a variable sky conditions otherwise one can use milli-voltmeter reading for instant values of TPAR.

The line quantum sensor is used at the top of canopy in inverted position to measure RPAR and directly above the canopy to get IPAR. Using all these values, one can estimate the APAR. This gives the good indication of PAR utilized by the crop. These readings taken over the crop period gives the utilization of PAR by the crop as it grows. The line quantum sensor readings vary with the particular aim or objective for which it is used. Since most of the time, relative measurements are required and the use of different units can be avoided by expressing in percentages. But when it is used for a specific purpose the units become very important.

16.1.3. Precautions

- The sensor should never to drop from a height.
- The sensor surface should be clean. It can be cleaned by moist cloth.
- It should be placed horizontally always other wise the readings will be changing.

Questions for Study

Q.1. What is PAR?

Q.2. How is line quantum sensor used in field?

Q.3. Write a note on line quantum sensor.

Q.4. What is the procedure to record PAR?

Q.5. Write the working principle and precaution of line quantum sensor.

❑❑❑

Chapter 17

Measurement of Crop Canopy Temperature

The most established method for detecting crop water stress remotely is through the measurement of a crop's surface temperature. The correlation between surface temperature and water stress is based on the assumption that as a crop transpires, the evaporated water cools the leaves below that of air temperature. As the crop becomes water stressed, transpiration will decrease, and thus the leaf temperature will increase. Other factors need to be accounted for in order to get a good measure of actual stress levels, but leaf temperature is one of the most important.

Canopy temperature and its reduction relative to ambient air temperature is an indication of how capable is transpiration in cooling the leaves under a demanding environmental load. The relationships between canopy temperature, air temperature and transpiration is not simple, involving atmospheric conditions (vapor pressure deficit, air temperature and wind velocity), soil (mainly available soil moisture) and plant (canopy size, canopy architecture and leaf adjustments to water deficit). These variables are considered when canopy temperature is used to develop the Crop Water Stress Index (CWSI) which is gaining use in scheduling irrigation in crops.

In plant breeding and selection for drought resistance, the interest is in finding genotypes that maintain lower canopy temperature as compared with other genotypes under the same field conditions. Relatively lower canopy temperature in drought stressed crop plants indicates a relatively better capacity for taking up soil moisture and for maintaining a relatively better plant water status. This capacity, as expressed in relatively lower canopy temperatures, was correlated with final yield under stress or other parameters like drought resistance in terms of yield such as various plant yield indices under stress.

17.1. Measuring Instrument

Canopy temperature is measured remotely by the Infrared Thermometer (IRT).

17.1.1. Infrared thermometer

It is a non-contact method for estimating the surface temperature of an object. It is widely used for measurement of crop canopy temperature and canopy minus air temperature (Tc-Ta), which are useful in detecting crop stresses, computing evapotranspiration and scheduling irrigation of crops.

In principle, the measurement does not interfere with the surface and yields a temperature that is an integrated value over the field of view of the sensor. Infrared thermometer (Model AG 42, TelaTemp Corp., Fullerton, CA; Fig. 59a, 59b & 59c) measures the radiation emitted from the object, and relates this radiation to the absolute surface temperature of the object by the Stefan-Boltzmann blackbody law.

$$R = e\sigma T^4$$

where,

R = radiation emitted,

e = Emissivity of the surface,

σ = Stefan-Boltzmann constant ($5.6/4 \times 10^{-8}$ $Wm^{-2}K^{-4}$)

T = Absolute temperature (^{o}K)

Most infrared thermometers have filters that allow energy in the 8-14 nm wave band to reach the sensor. This wave band includes the peak of blackbody emission at normal temperatures and has relatively low absorption by water vapour. The emissivity of the vegetation with in this wave band is usually high (0.94 to 1.00). Even with soil exposed between rows of trees/ plants, the composite emittance remains high. Considering the various factors involved, the error caused by assuming e = 1 for plant canopies would not generally exceed 1°C.

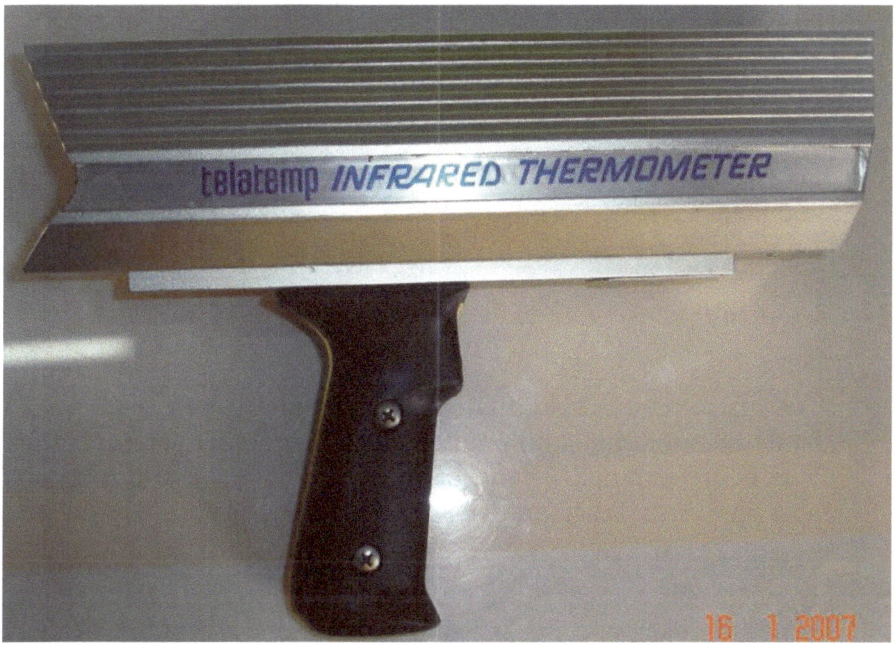

Fig. 59a: Infrared thermometer

17.1.2. Working principle

Thermal detectors convert the radiation into heat through a device which produces a signal proportional to the amount of radiant energy. In case of infrared thermometer, there is a concave reflector which converge all the radiation on to the sensor with a field of view of 4°. When the infrared thermometer is pointed towards the object, the emitted radiation falls parallel to the concave surface and converges at the focus where thermistor is kept, which is maintained at constant potential. The radiant energy gives out heat at the thermistor, which will reduce the resistance of the thermistor, which intern will reduce the current in the circuit. This current operates a digital circuit, which converts the current into temperature, using the emissivity values of the object.

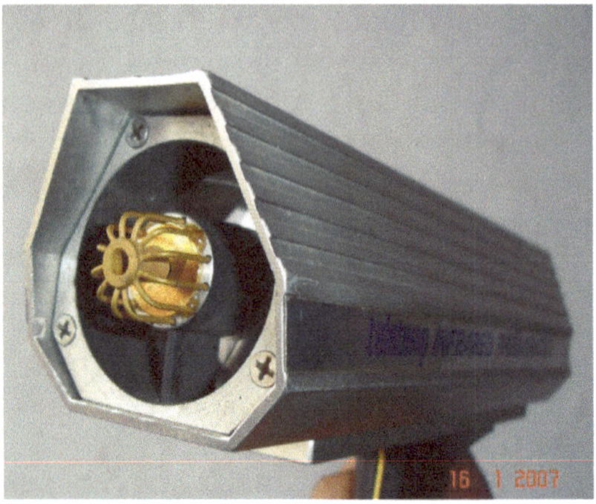

Fig. 59b: Front view of IR thermometer

Fig. 59c: Rear view of IR thermometer

17.1.3. Operating instructions

- The emissivity is kept at 0.95.
- The instrument is held by the hand grip at an angle of about 30° from the horizontal and pointed to the object whose temperature is to be determined. The instrument will stabilize in one second and give digital display.
- The digital display will give the average temperature of the entire scene which comes in the field of view. Normally the distance between object and the sensor is around one meter but to cover larger canopies, distance can be increased accordingly.
- By pressing the differential switch, the temperature difference between the object and the ambient air (Tc-Ta) can be obtained on the display.
- In the measurement of crop canopy temperatures, it is important to see that the soil back ground does not appear in the field of view of infrared thermometer.
- Canopy temperatures were taken as average of 8-10 measurements facing both east and west during 1200 to 1400 hrs to minimize sun angle effects.

17.1.4. Procedure

1. Charge the instrument well (overnight) before going to take the observation.

2. Take the instrument carefully to the field, where canopy temperatures are to be recorded.
3. Take out the instrument and hold it firmly by the grip. Point towards the crop canopy from a height of approximately 1 m, at an angle of 30° to the horizontal (Fig. 60).
4. Canopy temperature will be displayed. Note the reading.
5. Now press the differential switch to get the difference of canopy and air temperatures (Tc-Ta). Note this reading also.
6. Repeat the process from four or more directions of the plant canopy and each time take the readings.
7. Finally find the average of canopy temperature as well as that of difference of canopy and air temperatures (Tc-Ta).

Fig. 60: Observation in the field by Infrared thermometer

17.2. Advantages of infrared thermometer

- Physical contact with the measuring surface is not necessary.
- It has very rapid response time. The surface temperature can be obtained in one second.
- It is very rugged and portable.

- It is provided by rechargeable Ni-Cd batteries.
- The hand grip provides instant turn on-off the instrument.
- Accuracy is ±0.5°C.
- Operating environmental limit ranges from 20°C to 50°C.

17.3. Precautions

- Do not face the sensor towards the sun.
- Do not use the instrument with low battery warning.
- The instrument is to be charged regularly.

Questions for Study

Q.1. What is the importance of canopy temperature measurement in agriculture?

Q.2. Describe the working principle of infrared thermometer.

Q.3. Describe the procedure of observation in the field.

Q.4. What is the meaning of Tc-Ta?

Q.5. Write the precautions to be adopted during the working with infrared thermometer.

❑❑❑

Chapter 18

Measurement of Leaf Area Index

Leaf area is an important attribute in bio-productivity studies, as it bears an important and direct relationship to photosynthesis and transpiration. Leaf area is also important in determining the percentage of solar radiation intercepted by an individual plant or crop canopy. Photosynthesis centers on individual leaf. However, productivity depends not only on individual leaves but on stand of vegetation and canopy structure. The photosynthetic productivity of a canopy depends on both the amount of leaves present and the configuration of the leaves making up the canopy. Thus, leaf area has direct bearing on plant growth and final yield.

If we specifically wish to consider the productivity of crops or natural ecosystems, it is convenient to express their performance per unit land area. In field studies, the leaf area index forms the basis of the relationship between absorbed radiant flux and plant growth.

18.1. Leaf Area Index

Leaf Area Index (LAI) is defined as the one sided green leaf area per unit ground area in broadleaf canopies, or as the projected needle leaf area per unit ground area in needle canopies. The leaf area index (L) or leaf area per unit land area is computed as:

$$\textbf{Leaf area index (L)} = \frac{La}{P}$$

Where, La is the functional (green) leaf area of the crop canopy standing on ground area (P). As both La and P are normally expressed in m^2 (as areas), therefore, L is dimensionless. La may be determined by harvesting all leaf material above a quadrant marked out on the ground and it refers generally to the area of a single leaf surface.

18.2. Leaf Area Measurement

Many methods for measuring leaf area have been reported with varying degrees of accuracy e.g. relationship between leaf length or dry matter \ and leaf area, tracing/weighing procedures, dot counting, planimetric methods etc.

18.2.1. Manual estimation

Total leaf area for many plants can be derived by measuring length and breadth.

Leaf area (A) = k x L x B

Where,

L = length of leaf

B = breadth of leaf at mid point

k = constant determined by graph method for plant species under investigation

$$\textbf{Total leaf area} = \sum_{i}^{n} A$$

Where, n = number of leaves

18.2.2. Leaf area meter

The Licor Leaf Area Meter (LI-3000, Fig. 61) is designed for laboratory measurement of large and small leaves. Direct measurement of leaf area with an electronic area meter generally provides the lowest coefficient of variance and requires the fewest plants to be measured in order to detect comparable differences among crop canopies.

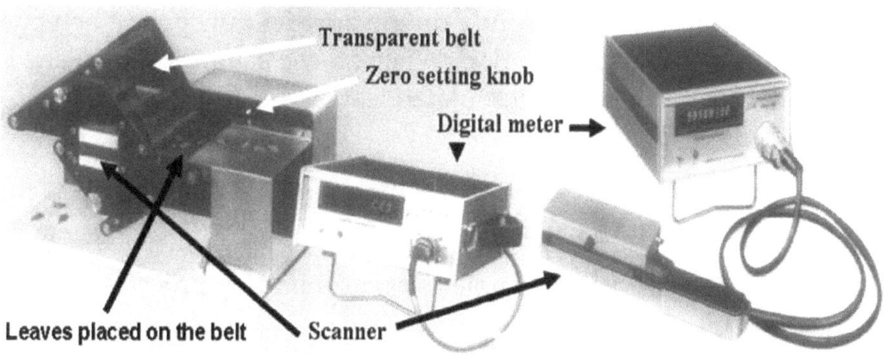

Fig. 61: Leaf area meter with their parts

18.2.2.1. Theory of operation

Leaves are placed between the guides on the lower transparent belt and then passed through the LI-3000. As the sample travels under the 15 watt fluorescent light source, the accumulating area in cm^2 is shown on LED display. The leaf image is reflected by a system of three mirrors to a solid state scanning camera within the rear housing. The mirrors are mounted on the instrument base plate. Two are visible under the transparent belts. The third is with in the rear housing. Individual leaves are placed on the belt which blocks the light and the leaf area is detected by blocked light and recorded by scanning camera. When no leaf is placed on the belt, no light is blocked, all the detectors receive light, and no area is recorded. With a stationary leaf positioned in the sensor, the area of blocked light is the area of the leaf (cm^2) for that leaf length. Holes in the leaf present no problem, for they will allow light to pass and will not contribute to the area. After recording the area of leaves, each time the belt is cleaned and whenever required. The belt is removed and thoroughly cleaned with liquid detergent.

18.2.2.2. Procedure

- First press the 'On' switch, a false reading will appear on the LED display and increases continuously. After pressing the lamp start switch, the false reading will be stable. Press the reset (zeroing) button for zero reading on LED display. If the LED display shows some readings, this may be due to the dust particles on the transparent belt. Clean the belt gently with the help of muslin cloth or tighten and adjust the belt in the middle until the LED display shows zero reading. After zero reading on LED display, the instrument is ready for measurement of leaf area.

- Place samples/leaves gently on the moving transparent belt one by one without any overlap. When all the samples are finished, note down the final accumulated leaf area reading which comes on LED display.
- After completion of one set of samples, press the reset button for zero reading on LED display.
- For more number of samples, repeat the same procedure again each time.
- After the measurement of leaf area of samples, calculate the leaf area index by using the prescribed formula.
- After completion of experiment, press the switch off button. The belt is then removed and thoroughly cleaned with the liquid detergent.

18.2.2.3. Calibration procedure

- Measure a paper of known area.
- Cut the paper into many small discs and take a reading.
- Adjust the "CAL" screw on front panel until the total area reading for all small discs equals the whole paper reading.
- The difference in the areas is an estimate of the inherent error due to a larger ratio of edge-to-surface area in the case of the cut paper.

18.2.2.4. Precautions

- Allow the instrument to warm up for at least 30 minutes prior to calibration.
- Check calibration frequently.
- Loose belt or stretched belt causes spurious counts. Tighten or replace belt.
- If fluorescent tube is not illuminated, Press "lamp start" switch.
- Calibration screw is too sensitive, turn the screw gently counters clockwise until displayed numbers remain constant.

Questions for Study

Q.1. Write the importance of leaf area measurement in crop production.

Q.2. How is leaf area index calculated?

Q.3. Describe the calibration procedure of leaf area meter.

Q.4. Write the working principle of leaf area index.

❑❑❑

Chapter 19

Measurement of Transpiration

Water loss from a plant in the form of vapour is known as transpiration. Loss of water vapour mainly from leaf cells through pores but also from the leaf cuticle and through lenticels of the stem. The resistance to diffusion of water vapour through a plant's stomata is called as stomatal resistance. Whereas, stomatal conductance is a numerical measure of the maximum rate of passage of either water vapour or carbon dioxide through the stomata, or small pores of the plant.

19.1. Measuring Instrument

The steady state porometer gives reliable measurement of stomatal resistance (Fig. 62). Its measurement is fundamental in studies involving photosynthesis, water and nutrient use, and drought tolerance.

19.1.1. Description of steady state porometer

1. *Display*: A liquid crystal display (LCD) exhibits the measured parameters and prompting displays.

2. *Read out control function switch*: This switch controls the display and is used when entering the aperture area and pressure.
3. *Aperture set potentiometer*: This is adjusted when entering the desired aperture area for diffusive resistance.
4. *Pressure set potentiometer*: This is adjusted when entering the desired atmospheric pressure.
5. *Null adjust value*: This is used for coarse adjustment of dry air flow entering the cuvette.
6. *Null adjust indicator*: "Null ADJ" meter indicates which direction to turn the null adjust valve in order to maintain the nulling humidity.
7. *Pre-dries desiccant pack*: This is one of the dessicant packs and will absorb the majority of the moisture in the dry air flow. A silica gel desiccant is used in desiccant pack.
8. *Desiccant IN*: Atmospheric air that has gone through the internal control flows out of this fixture into the dessicant pack.
9. *Desiccant out*: Air that has passed through the dessicant pack flows back in the instrument and into the sensor head hoseline.
10. *Sensor head cable connector*: The sensor head cable is connected and twist-locked into place before beginning measurements. The display will show: "Help" if the sensor is not connected or if the cuvette temperature drops below 0.4°C.
11. *Sensor head hoseling connection (to sensor):* The sensor head air hoseling is connected hereby screwing it into place.
12. Voltage reduction switch.
13. A/C power cord for recharging battery.
14. On and off switch.
15. *Data out put*: Various parameters *viz.*, 1. Aperture set, 2. Flow rate (cm^3 sec^{-1}), 3. CUV temperature (°C), 4., Leaf temperature (°C), 5. Relative humidity (%), 6. Quantum energy (mE m^{-2} s^{-1}), 7. Diffusive resistance (S cm^{-1}), 8. Transpiration (mg cm^{-2} sec^{-1}), 9. Pressure set- absolute atmospheric pressure (K Pa; where, 1 K Pa= 10 millibars).

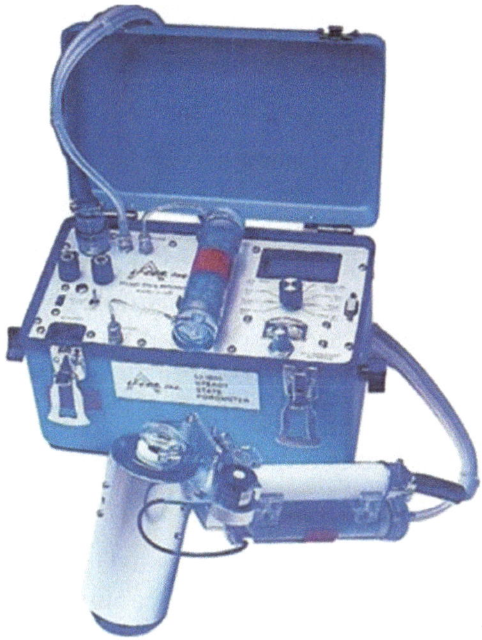

Fig. 62: Porometer

19.1.2. Working principle

Water are continuously being absorbed from the soil translocated through the plant and passed off into the atmosphere without ever being involved in any apparent function and also moves through the root system only when there is a driving force to move it. Water loss from a leaf is determined by maintaining a constant vapour density in a cuvette that is in contact with the transpiring leaf. This is achieved by pumping dry air into the cuvette at an appropriate measured rate to obtain a balance (at a predetermined humidity) between the flux of water transpired by the leaf and the flow of moist air out of the cuvette. Stomatal resistance is determined directly from the measured parameters. Steady state porometer works on the following principle.

The volumetric flow rate F ($cm^3\ s^{-1}$) of dry air into the cuvette can be expressed as

$$F = (Tc/273 + 1)\ (101.3/ñ)\ M$$

Where Tc (°C) is the cuvette temperature, ñ (kPa) is the absolute atmospheric pressure in the cuvette, and M ($cm^3\ s^{-1}$) is the mass flow rate of dry air into the cuvette as measured by the LI-I600 mass flow meter, referenced to standard conditions of 101.3 kPa, 0°C.

The transpiration T_r (mg cm^{-2} s^{-1}) of the sample is related to the volumetric flow rate by

$$Tr = (ñ_c - ñ_a)\ F/A$$

Where $ñ_c$ (mg cm^{-3}) is the water vapour density in the cuvette, equivalent to the product of the saturation vapour density at Tc and RH/100; A (cm^2) is the area of the sample, and $ñ_a$ ((mg cm^{-3}) is the water vapour density of the dry air stream entering the cuvette (LI-1600 assumes a constant 2% RH). If the sample covers entire aperture area, the read out will be Rs or Cs direct. If the area is unknown at the time of measurement, a normalized resistance can be measured and adjusted to the true value after the area is known.

The stomatal resistance Rs (s cm^{-1}) is then determined from

$$Rs = (\rho_1 - \rho_c) / Tr - Rb = A / F\ [(\rho_1 - \rho_c)/ (\rho c - \rho a)]$$

Where $ñ_1$ ((mg cm^{-3}) is the water vapour density in the leaf (the LI -1600 assumes the leaf saturated with vapour; T_{leaf} is also used in calculating $ñ_1$), and Rb (s cm^{-1}) is the boundary layer resistance. In this particular model, it takes Rb as 0.15 s cm^{-1}. The saturation vapor density vs temperature are stored in the system and are used in the calculation of $ñ_a$, $ñ_c$ and $ñ_1$. The stomatal conductance Cs (cm s^{-1}) is given by 1/Rs.

19.1.3. Measurement process

Initially the cuvette is acclimatized to the ambient conditions. The unit is operated for a few minutes with the null adjustment valve closed and the sensor head and leaf clamp in the open position. This will ensure that the cuvette has stabilized in ambient temperature and humidity. The cuvette should be acclimatized to conditions near the sample since humidity can vary near or away from the sample. A sample leaf *in-situ* is placed on to the cuvette aperture and held in place by clamping action (Fig. 63). A fully matured leaf (second or third from the top) without any insect holes or diseases on the surface is to be taken. Air flow through cuvette is allowed for 3 to 4 seconds and keeping knob in 'hold' position, the leaf sample is released from the aperture. The ideal time of observation is 10.30 A.M. to 11.30 A.M. as at this time the stomata are fully opened. The relative humidity-of the cuvette in contact with the sample is held constant. The diffusive resistance and water loss (transpiration) is determined by the instrument and indicated on the display. Repeat the measurement for at least four leaves around the same site/plot for each treatment.

Fig. 63: Measurement of transpiration with porometer

19.2. Precautions

- The desiccants should be dried in a oven after each use.
- The battery should be fully charged before use. Normally a battery will work only for 7 to 8 hours. Keep at least for 16-24 hours for charging it again.

19.3. Warning

- The **LI** -1600 should not be used in sulphur dioxide atmosphere as the humidity sensor can be easily damaged from acid laden dust or soot particles.
- Do not block the flow of air between the opening and the leaf.
- Adjust your body position suitably.

Questions for Study

Q.1. Define transpiration, stomatal conductance and stomatal resistance.

Q.2. List the different parts of porometer.

Q.3. Describe the working principle of porometer.

❑❑❑

Chapter 20

Measurement of Soil Moisture Content

The information on soil-water content is necessary from the point of view of efficient water management of crop production as water required by crops is to be met from the water stored in their root zone. Soil water content changes periodically as a result of changes in field water balance components namely, precipitation, irrigation, evapotranspiration, runoff and deep percolation. Crop growth is adversely affected if the soil water depletes below certain level. For selection of crops, deciding of irrigation level and frequency of irrigation, the assessment of water content of soil is important.

20.1. Soil moisture determination methods

Soil moisture content is estimated either directly by **gravimetric** (weight or volume basis) and **spirit burning method** or indirectly by **Tensiometer**, **Gypsum block** and **Neutron moisture meter.**

20.1.1. Direct measurement of soil moisture content

20.1.1.1. Gravimetric method

Working principle

It is the most commonly used method of soil measurement wherein loss of moisture is determined. In this method, weight of wet samples and oven dried samples are recorded. Soil moisture is expressed as percentage on dry basis either on weight basis or volume basis. When soil is stated to contain 10 % moisture on dry weight basis, it means 100 g of dry soil hold 10 g of water. When expressed on volume basis, it means 100 cubic feet of soil holds 10 cubic feet of water. The moisture percentage is generally expressed on dry weight basis unless stated otherwise.

Materials required

Soil sampling auger, moisture boxes, drying oven, weighing balance, metre rod and plastic sheet.

Sampling site

This varies with the growth stage and rows spacing of the crops. For instance, in case of wheat crop in between the rows whereas in case of wider spacing crops as in maize (a) within the row, (b) one-fourth width from the row.

Procedure

A levelled spot is selected in the middle of the field. In the cropped field, sampling should be done in between two plants in a row. The auger (Fig. 65) is driven to the desired depth and the soil sample is drawn and removes it from the groove of the auger on a plastic sheet. Mix the sample thoroughly. Then a composite sample of not less than 100 gram is taken in moisture box (Fig. 64). The box is covered immediately as considerable error can occur by evaporation of moisture if the soil samples remain exposed to atmosphere for long time. Cover the boxes with a cloth in the field to avoid heating due to insolation. Tipping off some soil near to the lid and weigh the soil sample (including moisture box) on a physical balance correct to the two decimals to get fresh weight of soil (Ww). The soil sample is dried in an oven at a constant temperature of 105 °C for 48-72 hrs to get a constant dry weight. After drying, the soil sample (including moisture box) are again weighed to get dry weight of soil sample (Wd).

Fig. 64: Moisture **Fig. 65:** Sampling auger

After deducting the moisture box weight from the Ww and Wd, the soil moisture is calculated by using following formula:

Soil moisture content on dry weight basis (%) = [(Ww- Wd)/ Wd] x 100

Where, Ww is weight of moist soil in gram and

Wd is weight of oven dried soil in gram.

Example: A fresh soil sample weighs 100 gram and after oven dry, it weighs 80 gram. Calculate the moisture content of the soil.

Solution

Soil moisture content on dry weight basis (%) = [(Ww-Wd)/ Wd] x 100

= [(100-80)/80] x 100 = **25 %**

Moisture content determined by gravimetric method on dry weight basis can also be expressed on volume basis by multiplying the soil moisture content (obtained by gravimetric method) with bulk density of soil.

Soil moisture (%) on volumetric basis (Pv)

= (Soil moisture (%) on dry weight basis) x (bulk density)

Example: If the soil moisture content on dry weight basis (Pw) is 11.1% and bulk density is 1.3 g/cc. Calculate moisture percentage on volume basis (Pv).

Solution

Pv = Pw x bulk density

= 11.1 x 1.3 = 14.4%

20.1.1.2. Volumetric Method

Moisture content on volumetric method basis can also be obtained by taking soil sample with core sampler (Fig. 66).

Fig. 66: Core soil sampler

Materials required

Sampling tube or core sampler, moisture boxes, weighing balance, drying oven, plastic sheet and metre rod.

Procedure

In this method, soil sample is drawn using a core sampler or a tube auger whose volume is known (V_1). The soil sample is transferred to an aluminum moisture cane or moisture box and the fresh weight is recorded (Ww). The soil sample is dried in an oven at a constant temperature of 105 °C for 48-72 hrs to get a constant dry weight. After drying, the soil sample is again weighed to get dry weight of soil sample (Wd). The soil moisture percentage is calculated by the relationship as given below;

Soil moisture (%) on volumetric basis (Pv)

$= [(Ww - Wd)/(Pw \times V_1)] \times 100$

Where,

Ww = wet weight or fresh weight of soil (g),

Wd = dry weight of soil (g),

Pw = density of water (g/cc i.e. 1.0 g/cc), and

V_1 = volume of soil including pores (cc)

i.e., $\pi r^2 h$ (volume of core).

Example: Soil sample is drawn with a core sampler having the dimension of 2.0 cm radius and 15 cm height. The fresh weight and dry weight of the soil are 591.8 g and 539.6 g, respectively. Calculate the moisture percentage on volume basis.

Solution

Volume of core = $\pi r^2 h$

$= (22/7) \times (2.0)^2 \times 15\ cm^3$

= **188.6 cm³**

Soil moisture (%) on volumetric basis (Pv)

$= [(Ww - Wd)/(Pw \times V_1)] \times 100$

$= [(591.8 - 539.6)/(1 \times 188.6)] \times 100$

= **27.7%**

Precautions

- As a general rule, the soil samples for use in determining moisture content should weigh 100 gram or more.
- About 10 samples per hectare seem to be reasonable for research work.
- In field experiment at least two replications of soil sampling should be done.
- Big size gravels should be avoided.

Limitation

It is laborious and also time consuming method.

20.1.1.3. Spirit burning method

Working principle

The principle in this method is to burn the soil moisture by igniting with spirit which is added in distinct units (installments). The process is repeated till a constant weight is obtained.

Materials required

Moisture boxes, glass rod, physical balance, measuring cylinder, spirit, sampling auger, etc.

Procedure

Weigh about 25-30 g of soil in a moisture box. Add 5.0 ml of spirit. Ignite, cool and record the weight. Add another lot of 5.0 ml of spirit and ignite, cool and record the weight. Repeat the procedure till constant weight obtained. Calculate the per cent of moisture on burnt dry weight basis.

This method is rapid and more suited for moisture determination where laboratory facilities are not available. However, burning method needs to be calibrated against oven dry method for each soil type.

20.1.2. Indirect measurement of soil moisture content

20.1.2.1. Tensiometer

It measures the soil moisture tension. This was first designed by Richards and Gardner in 1936. It consists of 7.5 cm long porous clay cup, a protective metallic tube, vacuum gauge and the hollow inner tube holding all the parts together (Fig. 67).

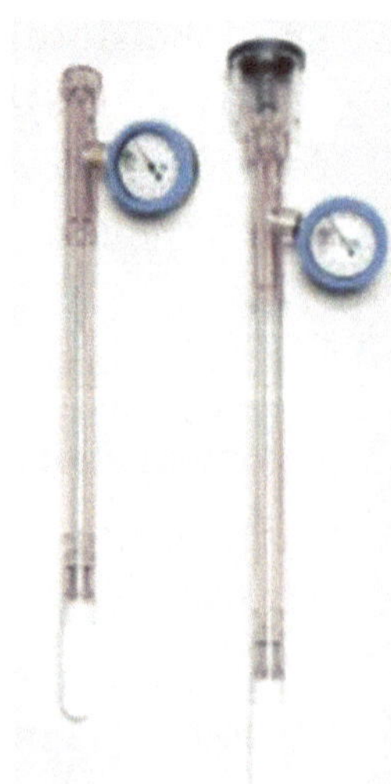

Fig. 67: Tensiometer

Working principle and procedure

Before installation, hole for the tensiometer is drilled using a gauge auger (Fig. 68). At the time of installation, the system is filled with water from the opening at the top and closed with rubber cork. When the system is installed in soil, moisture from cup moves out with the drying of the soil creating a vacuum inside the tube, which is measured with the help of gauge. When the desired tension is reached the crop is irrigated. Thus, Tensiometer is used for measuring the soil moisture tension at any time *in-situ* and to schedule irrigation accordingly. The porous cup should have high conductance and at the same time air should not leak in. For conductance to be high, the pores should as large as possible and must be freely permeable to water. If there is a leakage this sets a limitation in the use of Tensiometer and air leak value of a Tensiometer is generally about 1 (one) atmosphere. Hence, it should be used upto 0.85 atmosphere tension since it works satisfactory upto this much tension.

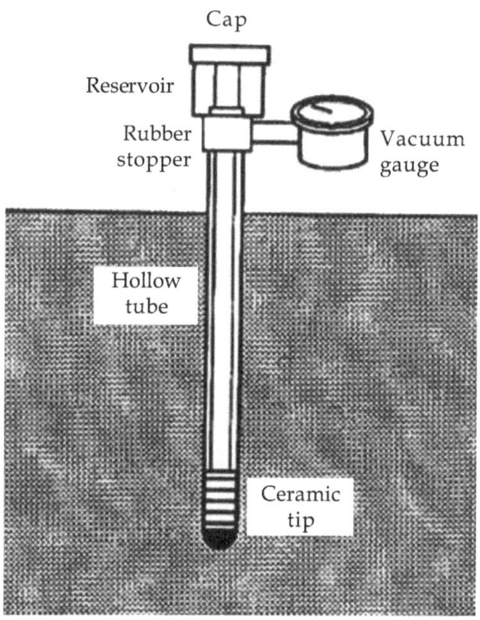

Fig. 68: Tensiometer installed in field

Precautions

- Use distilled water or boiled water which is free from gases for filling the Tensiometer.
- Do not hold the Tensiometer by its cup to prevent sealing of pores due to greasy palm.
- Install the Tensiometer in the active root zone of the crop.
- During installation ensure that there is a firm contact between the cup and the soil.

Checking of Tensiometer

Before installation, Tensiometer should be checked for its working. In order to find out whether the whole system is air tight, fill the tube with distilled water and cork it. Hold the Tensiometer cup above the hot plate. As the moisture evaporates from the cup surface, the needle shows increase tension reading till at least 0.85 atmosphere. If the system is defected the needle will come back before it reaches 0.85 atmosphere reading thereby showing the leakage.

Merits of Tensiometer

1. The instrument is very simple and easy to read the soil moisture tension *in-situ*.
2. It is very useful for scheduling irrigation to the crops which require frequent irrigation at low tension.

Demerits of Tensiometer

Sensitivity of a Tensiometer is only upto 0.85 atmosphere whereas, the available soil moisture range is upto 15 atmosphere. Hence, Tensiometer are more useful in sandy soil where in about 80 per cent available water is held within 0 to 0.85 atmospheric tension range.

20.1.2.2. Gypsum block or electrical resistance method

Gypsum blocks or plasters of paris resistance units are used for the measurement of soil moisture *in-situ* (Fig. 69). These were first designed by Bouyoucos and Mick in 1940 and since then considerable modifications and improvement have been suggested by various workers. The resistance blocks are made of different materials and gypsum blocks are most widely used.

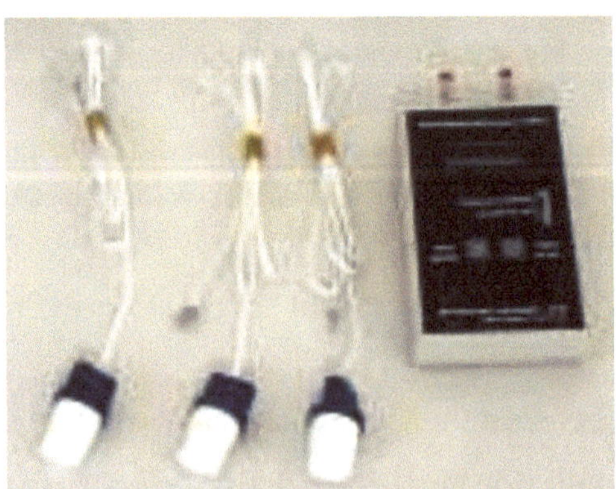

Fig. 69: Gypsum blocks

Working principle

It works on the principle of conductance of electricity or the principle based on the linear relationship between the electrical resistance and moisture content of the soil. Electrical resistance increases with decreases in moisture content.

Procedure

Two electrodes in the gypsum blocks are attached by wires to a resistance or conductance meter that measures the electrical resistance or conductance. Electrodes embedded with gypsum blocks are buried at an appropriate depth in the root zone at certain distance (Fig. 70). After being in soil for sometime, equilibrium is reached between that of surrounding soil. As the soil moisture increases or decreases, the water content of the blocks also increases or decreases correspondingly. The resistance offered to the flow of electricity depends on the wetness of the block. Higher the water content, higher the conductance or lower the electrical resistance and vice-versa. Thus, when the block is wet conductivity is high and the resistance is low. The actual resistance reading of block varies with the types of electrodes, their length and the distance between them. Generally, these read about 400-600 ohms resistance at field capacity and 50,000 to 75,000 ohms resistance at PWP. The readings are taken with a portable Bouyoucos moisture meter which is operating with a dry cell.

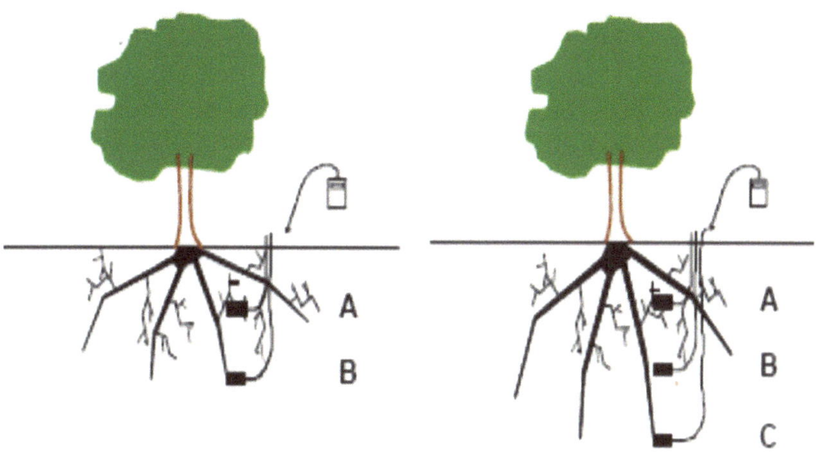

Fig. 70: Gypsum block installed at different depths in root zone

The relation between electrical resistance and soil water content/ tension is obtained in the form of calibration curve for each block.

This calibration curve is used to find out water content against the electrical resistance value when observed in respect of a block. Some resistance meters also show directly the available water present in the soil.

Properties of gypsum block

Gypsum has the property of expanding while setting. Thus, gives an excellent electro interface upon setting. The pores space of gypsum block is of such a size that capillary movement is facilitated between the block and the soil. Pure gypsum sets for about 30 minutes. The block lasts for about five years in a well drained soil but its life may be reduced to one season in the saline soil or high organic matters and wet soil. The blocks are resistive to soil moisture from 1 atmosphere tension to 20 atmosphere tensions. Therefore, the blocks are not reliable in the wet soils.

Limitations

1. This method is not suitable for soils at freezing temperature as it causes a marked increase in resistance.
2. This method can not be used for soils having high salinity as resistance drops with increase in salinity.

20.1.2.3. Neutron moisture meter

Neutron moisture meter is the most commercially used indirect method for measuring soil water content by emitting fast neutron from the source (Fig. 71). The phenomenon of the loss of maximum kinetic energy when neutron collides with hydrogen nuclei of water forms the basis of this method.

Working principle

The fast neutrons are emitted from the source (radium D) at a very high speed. When theses neutrons are collide with a small atom such as hydrogen contained in the soil water, their direction of movement is changed and slowed down. These slowed down neutrons are measured by using a metal foil composed of silver or boron tri-fluoride rendered radioactive when struck by the neutrons. The more the water (and thus hydrogen) present in the soil the larger will be the number of slowed neutrons which strike the metal foil and greater will be the resulting radioactivity which can be measured easily.

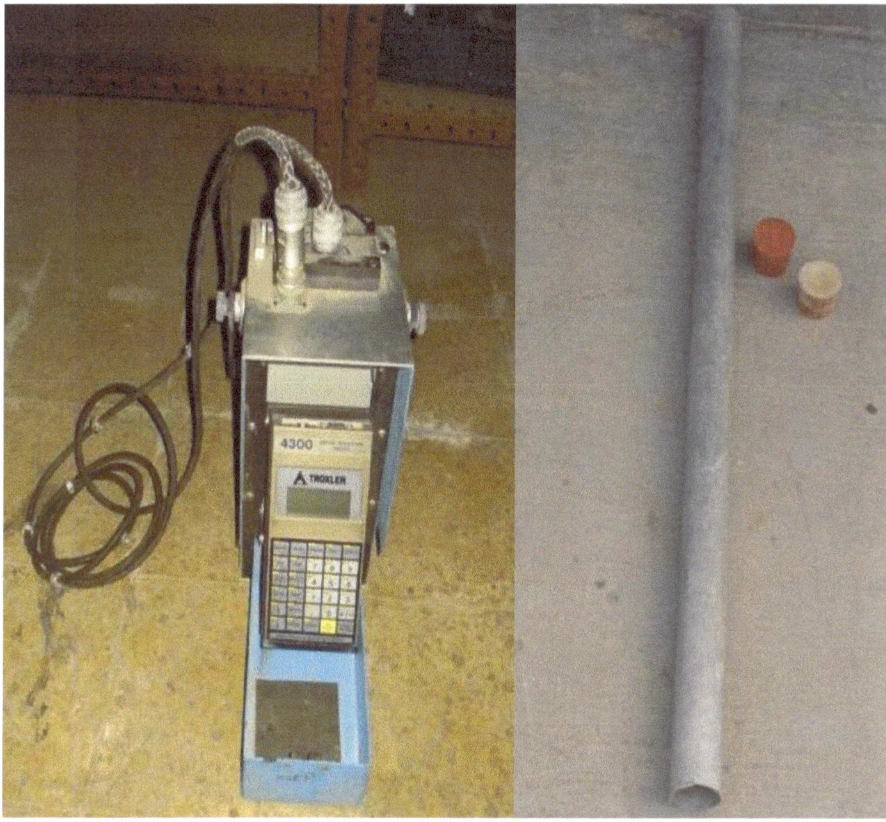

Neutron moisture meter   Aluminum access tubes

Fig. 71: Neutron moisture meter and aluminium access tube

Components of Neutron Moisture Meter: It consists of two parts namely

a) Probe and

b) Scaler or count recorder.

a) *Probe*: It contains a source of fast neutron 2-5 milli-curie mixture of Radium-Beryllium or 30 MC of Americium-Beryllium and a detector of slow neutrons (Fig. 72).

b) *Scaler or count recorder*: It is usually battery operated and portable (Fig. 72). It monitors the flux of slow neutrons which is proportional to soil water content (H^+). Aluminum access tube is not the part of the neutron moisture meter but required to place the probe into the access tube.

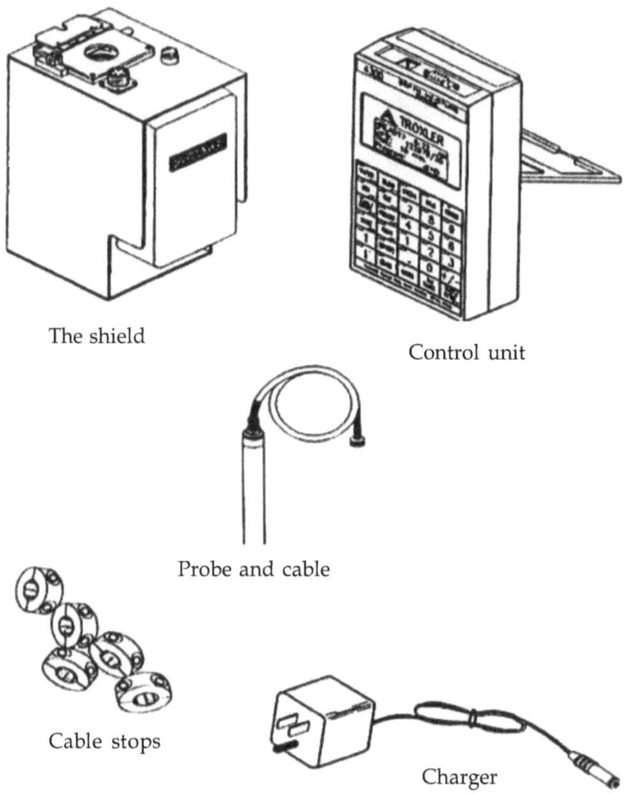

Fig. 72: Gauge parts and accessories

Determination of standard count

Prior to performing any tests, the standard count should be taken. The standard count will verify gauge operation. The gauge automatically compares the new standard count with the average of the last four standard counts. If the new count is within ± 2% of the last four average counts, the gauge will indicate "PASS".

Procedure

The Al access tubes are installed vertically into the soil. Place the neutron moisture meter over the access tube and the probe is lowered into the access tube (Fig. 73). The fast neutrons contained in the probe are emitted readily into the soil. They encounter and collide with various hydrogen nuclei and maximum loss of kinetic energy occurs with hydrogen nuclei which posses a mass nearly equal to that of neutrons. In practice, it has been observed that fast neutrons in soil are equal to the hydrogen content of the soil. The scaler or the rate meter monitors the flux of slow neutrons which can be converted into moisture content using a precisely developed calibration curve between water content and counting rate.

Fig. 73: Reading in the field with neutron moisture meter

Precautions

- Take full protection against radiation hazards. Follow instructions given by radiological safety officials. Wear radiation badges while operating the instrument.
- One should not put the probe near the soil surface as the neutrons which are radioactive in nature are not only hazardous to health but will also add error by their interaction or collusion with water vapour of the air. Probe must be 15 cm below the soil surface.
- Do not pull on the probe cable. It may damage the probe.
- The probe should remain in the shield except during measurement.
- The area of action of neutron moisture meter may be range from 10-30 cm depending upon the moisture content of soil.
- One should not use neutron moisture meter in the soil which is very rich in organic matter (H+ atoms).

- Calibration curve showing the relationship between count ratio and moisture content of each soil is essential. Count ratio = Actual count (measured count)/standard count. A range of data from very dry to very wet soil having 30-40 points is to be taken. This data is then feed to regression equations for the direct measurement of the water content.
- While taking count rate measurement, see that no other radio active source is lying nearby.
- Keep all tubes, at the same height above ground.
- The access tube must be kept dry.

Merits of the instrument

- It is a non destructive technique, allowing for long term studies on the same soil volume.
- Moisture content at any depth even in water logged and rocky soils can be measured, provided access tubes can be inserted.
- It is time saving as for a single measurement usually 2-5 minutes is enough and accurate method of determining soil moisture.
- The moisture content can be measured at exactly the same site throughout the season.
- A large volume of soil is analyzed in a single measurement as the process normally measures almost spherical volume with an average diameter of 35 cm to 75 cm is soil of very low soil moisture content.

Demerits of the instrument

- The initial cost of instrument is high. It suffers from radiation hazards.
- It requires technical competence.
- A special probe is needed for measurement to be performed on the top soil layer (15cm).

In spite of these limitations, the equipment is very useful for accurate and quick measurement of soil moisture in scientific studies.

Questions for Study

Q.1. Why soil moisture content determination is important?

Q.2. Write the merits and demerits of tensiometer.

Q.3. Why checking of tensiometer is necessary before installation in the field?

Q.4. What are the different components of Neutron Moisture Meter?

Q.5. Write the merits and demerits of Neutron Moisture Meter.

Q.6. Write the precautions while operating Neutron Moisture Meter in the field.

❑❑❑

Chapter 21

Measurement of Evapotranspiration

Evapotranspiration (ET) is the combined loss of water from vegetation and soil. When water is not a limiting factor at a site, the rate of ET is primarily controlled by meteorological factors like solar radiation, temperature, wind and relative humidity. The measurement of ET under natural conditions is of great importance in the water resource management and irrigation planning. Potential evapotranspiration (PET) is the evaporation from a surface if water was never a limiting factor. PE-can also be calculated using empirical formulae.

21.1. Measuring Instrument

The evapotranspiration is measured by means of lysimeters and it is the best method for ET estimation on day to day basis. Though the lysimeters provide precise and direct measurement of evapotranspiration, however the result is influenced by a major limitation/ assumption in reproduction of physical conditions such as temperature, moisture regime, density etc. within the lysimeter comparable to those outside it in the field. India Meteorological Department has installed both gravimetric and volumetric lysimeters for measurements of evapotranspiration at different stations in the country.

21.1.1. Measurement of evapotranspiration by Lysimeter

Lysimeters or evapotranspirometers are the instrument consisting of tanks filled with field soil and crops grown on it under natural conditions and used to measure evapotranspiration.

For measurement of ET, use of lysimeter is the best and convenient for obtaining ET data on a day-to-day basis. In lysimeter, the amount of moisture gained or lost by a suitably enclosed volume of natural soil is found through determination of the weight or volume of the soil mass at the beginning and at the end of a time interval. The lysimeter can measure either evaporation from soil surface or evapotranspiration from crops depending on whether the soil block and its surroundings are kept bare or are cultivated. There are two types of lysimeters:

1. Volumetric lysimeters which measures ET by volume.
2. Gravimetric lysimeters which measures ET by weight.

21.1.1.1. Volumetric lysimeter

In this instrument, the tank is connected by meant of under ground G.I. pipe to a reservoir through a float mechanism. The float mechanism controls the flow of water from the reservoir in such a way as to keep the water table in the tank to a certain height. Under this condition of water table, a known quantity of water is added and the excess water is allowed to percolate into a container (Fig. 74). The quantity of water added minus the percolated water gives the quantity of water supplied to raise the top layer of the tank to field capacity gives the ET loss. Rainfall, if any during the period is also added.

Fig. 74: Volumetric lysimeter

Working principle of volumetric lysimeter

The ET losses are measured by measuring volume of water evapotranspired during the desired time interval from a container. Loss of water due to ET causes water level in the soil tank to decrease which is compensated by means of the continuous supply of water to the soil tank from the reservoir. In this process, the level of water in the reservoir drum decreases. Then water is added to the reservoir drum daily around 8.30 A.M. to bring it to the height of the outflow tube fitted at the top of the reservoir. The quantity of water added is a measure of ET of the plants in the soil tanks.

Installation of volumetric lysimeter

The volumetric lysimeter consists of a soil tank, a float chamber and water reservoir. The soil tank is fabricated of mild steel sheet of 3 mm thickness having dimensions 120 x 120 x 90 cm with tapping towards bottom. The float chamber is constructed as basement of 90 x 60 x105 cm. The reservoir is an iron drum of cylindrical shape.

In the middle of the field, the soil tank is buried on a concrete base of 15 cm thickness under the ground at a depth of about 70 cm. The projection of the soil tank above soil level is about 20 cm. The tank is filled with soil up to the level of the field maintaining same soil profile as in the out side field.

The float chamber is also made in the ground on a concrete base of 15 cm thickness at a depth of 90 cm from the ground level with a projection of 20 cms above soil level. The reservoir is fixed above the ground on a concrete platform of approximately 45 cm height. The soil tank is in the middle of the field where as the float chamber and the reservoir are fixed on the border or the field. The reservoir is connected to the float chamber by means of a rubber tube. The float chamber is connected to the soil tank by means of a G.I. pipe laid under the earth at a depth of about 45 cms from the soil level.

The float unit is kept in the float chamber. It consists of a float fitted to a tripod stand. The height of the float can be increased or decreased by adjusting the screws fitted to the legs of the tripod stand. The float can be fixed at any desired height. The float consists of a very fine nozzle through which the flow of water is regulated at constant and steady rates. Thus, the water from the reservoir drum passes through a pipe into the float of lysimeter in the float chamber and through the G.I. pipe, it flows continuously into the soil tank maintaining a water table at a desired level.

Observation

The crop is grown in the soil tank along with the crop transplanted/planted in the field (Fig. 75). The level of water in the soil tank is maintained generally at a height of about 7 cm above the soil level for paddy and about 20 cm for other crops. Plant density, row and column spacing is maintained in the tank similar to the field. All cultural operations in the tank are performed similar to field with respect to time and manner etc. Some quantity of water in the tank is used by the plants for their transpiration purpose, whereas some water is evaporated from the water/ soil surface. The Total quantity of water used up for evaporation and transpiration purposes is the total quantity of water used for evaporation and transpiration purposes is the total quantity used for evapo-transpiration. Measurement of the evaporation can be obtained using the evaporimeter tank and daily ET can be found out. Weekly observations of crop are also taken to measure the height of the plants, tillers, panicles etc. and finally the yield of the plants of the soil tank is also calculated.

Method of Calculation

The ET is calculated as follows. On rain free days, the water added to the reservoir drum is measured with the 113 cc by volume and it is converted into millimeters by using conversion factor of 1 cc = 0.000694mm. On rainy days, outflow of water collected in the bucket in the float chamber is also measured. Inflow minus outflow gives net water used for ET purpose. This multiplied by the appropriate conversion factor as mentioned above gives as the net ET in mm.

Conversion Factor

Tank size = 1200 mm x 1200 mm.

Area 1200 x 1200 mm = 1440000 sq. mm= 14400 sq. cm.

Vol. of area in cc = 14400 sq. cm 1Q m = 10 mm. : . 1 cc = 0/14400 mm =1/1440 mm

Factor 1/1440 = 0.0006947

One for water added (30 divisions on a rain free day). The second on rainy day water added is 30 divisions and outflow is 20 divisions and the rainfall is say R mm. Examples for ET calculation are given below. Let the ET be Z mm.

1. 30 divisions = 30 x 113 = 3390 C.C.S.
   3390 CC = 3390 x 0.000694
   = X mm (say).

2. 20 divisions = 20 x 113 = 2260 CCS
   2260 CC = 2260 X 0.000694
   = Y mm (say).

The values are entered in the ET register as shown in Table 13.

Table 13 : Evapotranspiration register

Water Added (Divisions)	Inflow (1) CC	Outflow (2) CC	1-2 (3) CC	Rain (4) mm	ET (mm) *	EP (mm)
1) 30	30 x113 =3390 CC	NIL	3390 CC	NIL	X	Z
2) 30	3030x113 =3390 CC	20x113 =2260 CC	1130 CC	R mm Say	(x-y)+R mm	Z

During heavy rains, water over flow and ET value is not calculated and overflow is entered in the ET register. Generally on rain free days water added to the reservoir drum is a measure of ET. But on rainy days, there may be back water flow from the soil tank into the float champers. This back water should be measured and appropriate correction for rain should be applied to the ET algebraically as expressed above. On very heavy rainy days, such as in cyclones, the float chamber and the soil tank is flooded. Observations are suspended on these days and after the cessation of heavy rain, the float chamber is to be drained, cleaned along with the lysimeter. The level of water in the soil tank is brought to the constant level by removing water in the tank without causing any damage to the plants in the tank. Test observations are taken and regular observations are resumed as early as possible.

Fig. 75: Volumetric lysimeter

Precaution

- Care is to be taken that the water in the reservoir drum is clean and there are no impurities in the water.
- Keep the float clean by removing the impurities in the nozzle.
- During crop free seasons, the lysimeter, reservoir, float chamber and soil tank may be got painted annually.

21.1.1.2. Gravimetric lysimeter

Gravimetric lysimeter consists of a sensitive dormant type weighing machine of two tons capacity. It is erected in the middle of the field on a reinforced concrete foundation. The steel tank carries a perforated plate at a depth of 75 cm, so as to form a hallow chamber at the bottom. A tap is also fitted at the base to drain out the percolated water from the hallow position of inner tank. A tube inserted through perforated sheet into the bottom of the hallow chamber facilitates the removal of percolated water. Weight of the lysimeter is observed every day a 0700 hrs LMT. All crops except paddy can be grown in the gravimetric lysimeter (Fig. 76).

Fig. 76: Gravimetric lysimeter

Construction of gravimetric lysimeter

A weighing lysimeter consists of the following parts:

1. Retaining tank

This is the outer and bigger tank of size 14 x 14 x 11.2 m^3 which isolates the lysimeter and weighing bridge from the main field. It protects the steel tank in which crop is grown, from the water seepage from the adjoining field.

2. Steel Tank

It is inner tank of size 1.3 x 1.3 x 0.9 m^3 and fitted with retaining tank and connected to a weighing machine. It contains soil in which the plants are grown and its rim is in the level with surrounding soil surface. A perforated plate is fitted at a depth of 0.75 m in the steel tank to form a hollow chamber. A tube is inserted through the perforated plate to remove the percolated water from the tank.

3. Weighing machine

The weighing machine is a simple weight machine and has capacity to weigh upto 2000 Kg (Fig. 76). The weighing machine can be read correct to 200 gm i.e., the accuracy in terms of depth units of 0.12 mm of water.

4. Dummy tank

To prevent the over heating of steel tank side through gap between steel tank and weighing machine, a smaller tank of the size of 0.3 x 0.3 x 0.9 m is placed in this gap. This tank is known as dummy tank which is not used for ET measurement.

Working principle of gravimetric lysimeter

The working is based on simple weighing machine. The loss in weight during at fixed time interval gives the ET values.

Installation of gravimetric lysimeter

A concrete foundation is laid out in a pit size of 3.5 x 3 x 2 cubic metres. A steel tank (1.3 x 1.3 x 0.9 cubic meters) is mounted on the platform of the machine such that its rim is in level with surrounding soil surface.

Weighing machine is fitted on a platform of size 1.2 x 1.2 m^2. It is erected in the middle of a field on a reinforced concrete foundation, constructed inside a pit of size of about 3.5 x 2 m^2. When in position, the platform of the weighing machine will be at a depth of about 1.0 m from the soil surface.

Time of observations

Lysimeter observation is to be recorded at 0700 hrs LMT. Take the reading when the wind is almost calm, so that the correct weight loss is recorded. The change in weights (kgs) of tanks during the period between two concurrent observations should be multiplied by a conversion factor of 0.6 and corrected for rainfall during the same period gives evapotranspiration in mm.

Source of Error

Water logging in the foundation or in the false bottom of the field tank can lead to erratic readings. Also the movement of the balance should be absolutely free and its readings should be taken in calm or very light wind.

Care of Field and Crop

The crop in the tank and the field should be exactly similar in plant density, crop geometry etc. If a delay in germination is noticed either in the lysimeters or adjacent areas, they may be lightly watered. There should not be any un-cropped periphery around the lysimeter. Lysimeters observation may be started only after seedlings (clearly seen) except in case of sugarcane. If any dropping or withering of plants due to water stress in the tank is seen, the same may be fully drenched with water at the time of sunset. The weights of the tank before and after watering should be recorded. The 24 hours ET is computed by adding ET loss from morning to evening before watering and from evening after watering to the next morning.

Precautions

- Check and correct any water logging in the foundation or bottom of field tank.
- Ensure free movement of the balance.
- Take weight when wind is calm or light.
- Crop density in the tank and field must be same.
- Sensitivity of the balance regularly checked with the help of a known small weight.
- The outlets of the tank may be closed tightly to avoid loss of weight due to outflow.
- Annual overhauling of the entire lysimeter is done.

Method of ET Calculation

ET (mm) = Difference in weight of two consecutive lysimeter observations in kg x 0.6 + rainfall (mm)

If the rainfall is more then the rate of ET than ET equal to rainfall minus difference in weight x 0.6. If tank is over flowed, do not record the ET but mentioned overflow. Take the initial reading for next day calculation of ET.

***Exercise*:** Calculate the evapotranspiration (ET) by lysimeter using the following data:

Weight on first day at 0700 hrs	: 1763.4 kg
Weight on second day at 0700 hrs	: 1762.1 kg
Rainfall	: 4.1 mm

Solution

The evapotranspiration (ET) by lysimeter may be computed as:

ET = ΔW x 0.6 + rainfall

Where, ΔW = Change in weight of lysimeter (kg)

Change in weight of lysimeter = 1763.4 - 1762.1 = 1.3 kg

Rainfall = 4.1 mm

Therefore, ET = 1.3 x 0.6 + 4.1 = **4.88 mm/day**

Questions for Study

Q.1. What is potential evapotranspiration?

Q.2. What are the common methods of determining the evapotranspiration?

Q.3. How is evapotranspiration determined?

❑❑❑

Chapter 22

Study of Automatic-weather Station

Real time weather data at different time intervals is very important for the crop weather studies, but through the agro-meteorological observatories such real time weather data are generally not collected. The acquisition of agro-meteorological data through conventional observatories requires services of trained personnel to collect, compile and creation of weather data base. With the advances in technology, weather observations are automatically recorded, stored and can be transmitted to the remote place with the help of Automatic Weather Station (AWS).

22.1. Automatic Weather Station

An automatic weather station (AWS) is an automated version of the conventional weather station to enable measurements from remote areas or data collection platforms (DCP). The specific configuration may vary due to the purpose of the system (Fig. 77). The system may report in near real time via the Argos System and the Global Telecommunications System. The communication may be made through land line, radio or satellite link. AWS comprises of a field unit fitted with electronic sensors and data loggers of varying capacity for data storage. The system is powered by a battery which is continuously recharged by a solar panel. These systems can be either directly connected to a computer or may be fitted with a communication device to transmit the data to a distant user. The field unit is normally located in a fenced plot measuring 7.5 m x 6.0 m or 10m x 10m having prescribed exposure conditions for the sensors laid down by national and international meteorological organizations.

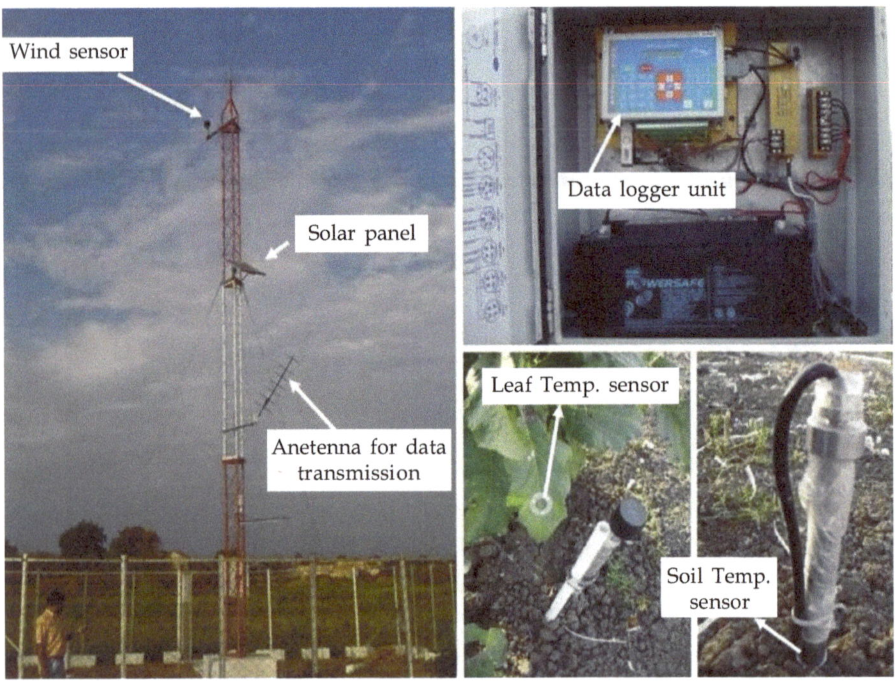

Fig. 77: Automatic weather station with different sensors

The automatic weather station consists of different sensors for measurement of real time weather data. The sensors measure rainfall, solar radiation parameters, pressure, air temperature, leaf temperature, leaf wetness, humidity, wind speed and wind direction. Except the pressure sensor and the rain gauge, the other sensors are mounted on a 3 m tower.

22.1.1. Sensors

Following are different sensors of an AWS used for recording observation of different meteorological elements at real time basis:

1. Sensor for measurement of air temperature used in AWS is platinum resistance thermometer. The sensor is a thermistor which functions on the principle of variation in the resistance with variation in the air temperature. The sensor has fast response time and has accuracy to measure 0.1 degree celsius temperature.
2. A tipping bucket rain gauge provides rain rate information with a resolution of 0.5 mm. It consisting of a twin bucket, whose collecting area bears a definite relation with the rainfall, mounted on a horizontal shaft which has two stable positions is used. The weight of rainwater causes tilting of the bucket to one stable position and brings the second bucket under the funnel for collecting water. Every time the bucket tilts, a small magnet attached to the horizontal shaft of the bucket actuates a red switch generating one electrical pulse.
3. The atmospheric pressure is measured through transducer consisting of a special temperature compensated aneroid sensor. It is made up with temperature compensated piezo-resistive pressure sensor, which is sensitive to pressure variation and measure the pressure with a resolution of 0.1 mb.. The sensing element is mechanically coupled to a micro-torque potentiometer. The expansion/contraction of the aneroid with pressure changes causes linear variation in the output resistance value of the potentiometer.
4. Three cup rotor type sensor is used to measure wind speed. The sensor is an AC voltage generator which uses a six pole magnet to rotate and a suitable stator coil winding. The frame carrying the three equally spaced cups rotates the anemometer rotor shaft with the wind, thereby including AC voltage in the stator coils. AC voltage output of the anemometer generator varies linearly with wind speed.
5. Potentiometer type sensor is used to measure the wind direction with a resolution of 1degree. The potentiometer is coupled to the wind vane shaft and gives a resistance output linearly proportional to the wind direction from 0° to 360°.
6. Thin film capacitance sensor is used for humidity measurement. The sensor used for measuring relative humidity is a thin film capacitor which is a chemically treated polymer thin film acting as a dielectric of a variable capacitor between two metalised electrodes. The absorption of moisture by the thin film changes

its dielectric constant thereby varying the capacitor value. The sensor with its associated electronics is housed in a small compact metallic tube. The output of the sensor varies from 0 to 1.0 v for 0 to 100% humidity and has accuracy of 3 per cent.

7. The sensor used for the measurement of sunshine duration is a sunshine-switch consisting of six tiny solar cells connected in series for obtaining sufficient output for actuating a solid state relay when there is sunshine.
8. For measurement of global radiation, the sensor consists of thermopiles as the receiving surface and an optical hemispherical glass dome for filtering out the terrestrial radiation component.
9. The Net Pyrradiometer measures the net total radiation flux (solar, terrestrial, and atmospheric) downward and upward through a horizontal surface.
10. Leaf temperature sensor is used for measurement of leaf temperature. It is a digital interface to allow connection of thermistor temperature sensors to the Logger. It provides a means of connecting up to four thermistors of the specified type to the system.
11. For measurement of Leaf Wetness the Dielectric Leaf Wetness Sensor (LWS) is used. The Leaf Wetness Sensor determines the presence and duration of wetness on a leaf's surface.

22.1.2. Power supply

The main power source for an automatic weather station is usually one or more solar panels connected in parallel with a regulator and one or more rechargeable batteries. As a rule of thumb, solar output is at its optimum for only 5 hours each day. In the Northern Hemisphere, the solar panel would be mounted facing south and vice versa for the Southern Hemisphere. The output from the solar panels may be supplemented by a wind turbine to provide power during periods of poor sunlight.

22.1.3. Data conversion, storage and transmission sub-system (DCSTSS)

The DCSTSS mainly consists of digital unit, transmitter and antenna. The system automatically takes the meteorological observations at specified times, stores them in the system memory and transmits in a self timed pseudo-random manner in its prescribed time slot within the next 60 minutes before the next observation takes place.

22.1.4. Precautions

- The 3 meter mast is used for the measurement of parameters that effect and it height is referenced to crop top.
- The 10 meter mast is used for the measurement of parameters such as wind speed and direction without interference from objects such as trees, buildings or other obstructions.
- To take measurements of wind, humidity and temperature, sensors are mounted at 30, 10, and 2 meters respectively. Other sensors are mounted around the 2 meter or lower height.

Questions for Study

Q.1. What is the significance of hourly weather data in agrometeorology?

Q.2. What is type of sensor is used for temperature measurement in AWS?

Q.3. Why tipping bucket raingauge is used in AWS for rainfall measurement?

Q.4. Write the precautions to be adopted for AWS.

Q.5. Write a note on sensors used in AWS.

❏❏❏

Chapter 23

Measurement of Upper-air Observation

Weather phenomena are not localized, it moves and varies over space and time. The weather over a place or area is the result of far away weather events/conditions. Acquisition of global scale observation, therefore, becomes a pre-requisite for understanding the physics of the weather and also to represent the weather condition over an area at a particular time. Complex three-dimensional models of weather systems can be made by collecting weather data at multiple levels in the atmosphere. Upper-air data are important to know present weather over a place and crucial for weather forecasting.

Surface weather data measurements are important to meteorologists, agriculturist hydrologists and planners. The surface weather data are measured at a place knows as weather station or meteorological observatory and the observations give a snapshot of a variety of weather conditions at one single location. These weather stations/observatories are installed on land, mountain and a ship to take the data from different locations on earth surface. The measurements taken at a weather station can include any number of atmospheric observables like temperature, pressure, wind, humidity and rainfall measurements, and so on so forth. To represent the atmospheric conditions at a location, we need weather data at different levels to represent it at three dimensional scales. Remote sensing makes it possible to collect weather data from remote or inaccessible area. This technique is used in weather satellite, radar, weather buoy sonar etc. to collect weather data. In earlier chapters, we have studied about the data acquisition from a surface weather station, but the surface data is also collected from a ship or a weather buoy and they are presented below:

23.1. Floating buoys and Marine surface observatories

Floating buoys carry surface meteorological instruments and provide weather data over sparse oceanic regions. Observatories stationed on ships are recorded marine surface meteorological data (Fig. 78).

Fig. 78: Meteorological instruments on ship

23.1.1. Observations by Buoys and Floats

IMD deploys drifting ocean data buoys in the vicinity of ocean and sea to make autonomous observations of atmospheric pressure, sea waves and water temperature while floating on the sea surface. In the area of sub-surface observations by floats, Argo floats are deployed to observe water temperature and salinity from the sea surface to a depth of 2,000 meters, and the data obtained is automatically transmitted via satellite.

The vessels also observe elements relevant to the marine environment such as heavy metals, pollutants and concentrations of carbon dioxide and other greenhouse gases in seawater and in the air over the sea. IMD receives marine meteorological reports from merchant and fishery vessels on a regular basis.

23.2. Pilot Balloon/ Meteorological Balloons

A Pilot Balloon is a meteorological balloon that is filled with gas (hydrogen or helium) lighter than air. When the pilot balloon is used in conjunction with a theodolite, it is used to determine the speed and direction of winds at different levels of the atmosphere.

It is a large balloon which used to collect upper-air observation. It is released for measurement of wind speed and direction and other weather elements at different height in the atmosphere using a theodolite. The balloons carry transmitting equipments for collection of upper air meteorological data and for determining the heights of the cloud bases also. To measure temperature, humidity and pressure at different heights in the atmosphere, meteorographs are attached to the ascending balloons (Fig. 79). The weather data recorded on the meteorographs is collected, when it falls from balloons and come on the ground. Some times when meteorographs are not found data pertaining to humidity, temperature and pressure are not recorded. The collected data are then plotted in weather map for analysis of weather situation at a particular time, which is further used for forecasting purpose. The wind speed and direction at selected levels are calculated either by trigonometric methods or graphical methods.

Fig. 79: Pilot balloons

23.2.1. Description

The balloons are in spherical, expansible type and their chief requirement, apart from being able to reach satisfactory heights, is that they should be capable of ascending at a constant rate very near to pre-determined value, which requires that they maintain a spherical shape even after expansion due to being exposed to reducing atmospheric pressure throughout their ascent (Fig. 79). Night observation is accomplished by attaching a lit paper lantern to the balloon. The balloons sounding stations are established by IMD in India to collect the upper air weather data.

23.2.2. Theodolite

The theodolite is an optical telescope, which measures horizontal and vertical angles of the ascending balloons at known intervals of time. It consists of a sighting telescope mounted in manner so that it is free to rotate around a horizontal and a vertical axis and has graduated scales so that the angles of rotation may-be measured while tracking the pilot balloon. When sky is obscured, it is difficult to tract the balloon through theodolite. The radio-theodolites are used which use radio signals to track the balloons.

Remote sensing techniques is of active type and is used for meteorological data collection. Remote sensing is a concept used to monitor remote weather events and subsequently, producing weather information. The common types of instruments used for meteorological data collection using remote sensing techniques are Radar, Lidar, and Satellites.

Upper-air observations are performed by two different networks: the radiosonde observation network and the wind profiler observation network.

23.3. Radiosonde

The radiosonde is a device which has sensing sensors on board and transmits observation by radio signals to a ground receiver as the balloon ascends. The instrument measures temperature, humidity and pressure at different height of the atmosphere through which it passes. The instrument is carried aloft by a meteorological balloon inflated with hydrogen and also mounted on board with aircrafts. It is carried aloft also by means of a free balloon. The radiosonde has a built-in high frequency transmitter that transmits data from the radiosonde meter and recorded on the ground by a specially designed radiosonde receiver. The radiosonde covers the upper air to a maximum of 30 km from the surface and measures temperature, pressure, humidity and wind. Radiosonde observations are carried out twice a day at 0000 and 1200 UTC.

The advantage of radiosonde is that the recorded data are immediately available and useful in weather analysis and forecasting.

23.4. Satellite

Satellite observations provide a wealth of information including data on the height and distribution of clouds, upper-air wind, and sea surface temperature distribution. The observational data received from the spacecraft and other National Meteorological and Hydrological Services to continuously monitor significant meteorological phenomena such as cyclones, fronts and low pressure systems, and is also directly assimilated into models for numerical predictions.

Satellite is a device circling the earth, watch the remote objects and collect data from remote sensing devices flown onboard satellites (Fig. 80). Satellites make measurements indirectly by sensing electromagnetic radiations coming from the surfaces below. There are two types of weather satellites *viz.*,

1. Geostationary and
2. Polar orbiting satellites.

Fig. 80: Satellite

Cloud pictures, its motion and development, solar radiation, cloud temperature are recorded from geostationary satellite. Temperature and moisture sounding, liquid water estimates, rainfall rates and humidity are recorded through polar orbiting satellite.

23.4.1. Geostationary satellite

Geostationary satellites are positioned at an height above the earth about 36000 Km. At this height, they rotate around the earth at the same speed as the earth rotates around its own axis, so that it remains stationary above a point on the earth (normally directly overhead the equator). As they remain stationary, they are ideal for use as communications satellites and also for remote imaging as they can repeatedly scan the same points on the earth beneath them and thus is able to capture the whole Earth disc at once.

In India, Kalpana-1 located at Longitude 74° E and Insat-3A located at 93.5° E, both geostationary satellites are being used to monitor the weather of Indian region.

Some of the parameters which are derived from INSAT -3D sounder payload are given below:

1. Temperature and humidity profile using brightness temperatures.
2. Geo-potential height using sounder retrieved temperature and humidity profiles.
3. Layer perceptible water using retrieved humidity at standard pressure levels.
4. Wind index using geo-potential height and retrieved temperature and humidity profiles at standard pressure levels.
5. Ozone using brightness temperatures.

23.4.2. Polar satellite

Polar satellites have a much lower orbit, moving around the earth fairly rapidly and scanning different areas of the earth at relatively infrequent periods. A polar orbiting satellite circles the earth at a near-polar inclination, meaning that it always passes almost exactly above the poles. The satellite passes the equator and each latitude at the same local solar time each day, meaning the satellite passes overhead at essentially the same solar time throughout all seasons of the year. The low earth orbit (800 - 850 kilometers) is much closer to earth than a geostationary orbit, and thus can see a smaller part of the earth than a geostationary satellite, but in finer detail.

23.5. Radar

It is based on radio detection and ranging, which make it possible to track clouds and precipitation at a distance of 300 -500 km (Fig. 81). Radar can be used for locating clouds, liquid water in a cloud, cloud depth, tracking of upper winds, cyclone, thunderstorm and other weather phenomena. It can also provide a synthetic display of the cloud and also distinguish water droplets, ice particle and their growth.

Different types of radar (radio detection and ranging) are used for meteorological data collection.

Fig. 81: Radar

23.5.1. Doppler radar

Doppler radars can measure not only precipitation intensity but also the radial velocity of raindrops. Doppler radar has additional capability of detecting motion in the atmosphere and is used for observe thunderstorm and cyclone development or even turbulence above airport runways.

Conventional radar provides information only on reflectivity, whereas, Doppler Weather Radars provide information on velocity and spectral width in addition to reflectivity.

23.5.2. Wind finding radar

Wind finding radar determines the speed and direction of winds aloft by means of radar echoes. A radar target is attached to a balloon and this target is tracked by ground radar. The bearing and time interval of the echoes is evaluated by a receiver.

23.5.3. Weather surveillance radar

A Weather surveillance radar is of the long range type which detects and tracks typhoons and cloud masses at distance of 400 kilometers or less. This radar has a rotating antenna disk preferably mounted on top of a building free from any physical obstruction (Fig. 82). Radio energy emitted by the transmitter and focused by the antenna shoots outward through the atmosphere in a narrow beam. The cloud mass reflects a small fraction of the energy back to the antenna. This reflected energy is amplified and displayed visually on a radar scope. The distance or slant range of the target from the radar is determined through the elapsed time, the signal is transmitted and then received as an echo. Its direction is determined by the direction at which the focused beam is pointing at the instant the echo is received. The radar is a useful tool in tracking and monitoring tropical cyclones.

A network of conventional Cyclone Detection Radars (CDRs) has been established at Kolkata, Paradip, Visakhapatnam, Machilipatnam, Chennai and Karaikal along the east coast and Goa, Cochin, Mumbai and Bhuj along the west coast. The Doppler Weather Radars (DWRs) have already been installed and made operational at Chennai, Kolkata, Visakhapatnam and Machlipatnam.

While conventional weather radar can look deeper into a weather system to provide information on intensity rain-rate, vertical extent, the capability to probe internal motion of the hydrometers and hence, to derive information on velocity and turbulence structure has become available only with the advent of Doppler Weather Radars (DWRs) which provide vital information on radial velocity from which wind field of a tropical disturbance in the reconnaissance area of DWR can be derived. In addition to above, a number of derived parameters useful for cyclone monitoring and prediction are also available from DWR.

India Meteorological Department, at present has 14 S-band and 25 X band radar and 5 DWRs radar network.

Fig. 82: Weather Surveillance Radar

23.6. Lidar

It use the laser beam and employs optical technique to determine cloud characteristics. It can also detect solid particles and tract their movement in air.

23.7. Sodar

It is an instrument, which uses acoustic echo sounder and depends on the backscattering of sound waves from air parcels having different density. It is useful for identify the intensity of disturbances in lower tropospheric layer.

23.8. GPS in meteorology

The Global Positioning System (GPS) is a satellite-based navigation system made up of a network of 24 satellites placed into orbit in six orbital planes at an altitude of 20000 Km above the earth surface. GPS works in any weather conditions, anywhere in the world, 24 hours a day.

23.8.1. Use of GPS in meteorology

- Nowcasting (for example Fog forecast, Flash flood monitoring, Thunderstorm etc.).
- NWP model for short term forecast between 6-12 hours.
- Climate studies.
- Integrated Perceptible Water Vapour (IPWV) comparison with other independent method.
- *Integrated Perceptible Water Vapour (IPWV):* It is The amount of atmospheric water vapour (in kilogram) overlying per unit area of the earth surface. Its unit is kg/m^2.

23.9. Synoptic observations

Simultaneous observations of the state of the atmosphere are made at a large number of locations all over the world several times a day. Surface observations made every 3 hours and upper-air observations every 6 hours. The density and frequency of the observations depend on the requirements as well as the technical and financial capability of the concerned country. These observations are referred to as synoptic observations and the time as synoptic hours. Synoptic observations are made all over the world at 000 GMT and thereafter every 3 hours interval round the clock on all 365 days of the year.

In India, the main surface synoptic hours are 0300 (830 IST) and 1200 (1730 IST) hour GMT. Other important synoptic hours are 006 and 1800 GMT.

23.9.1. Synoptic charts or weather map

A weather map is a tool used to display information quickly, showing the analysis of various meteorological quantities at various levels of the atmosphere. They have been in use since the mid-19th century and have been used both for research and weather forecasting purposes. Weather observations made at a specified hour at a number of stations can be plotted on a blank weather map. The plotted map, called the weather charts, gives a picture of simultaneous weather conditions over a vast area. Such a chart is also called the synoptic weather chart (Fig. 83). Analysis of a sequence of synoptic weather charts provide us information on weather development and its movement.

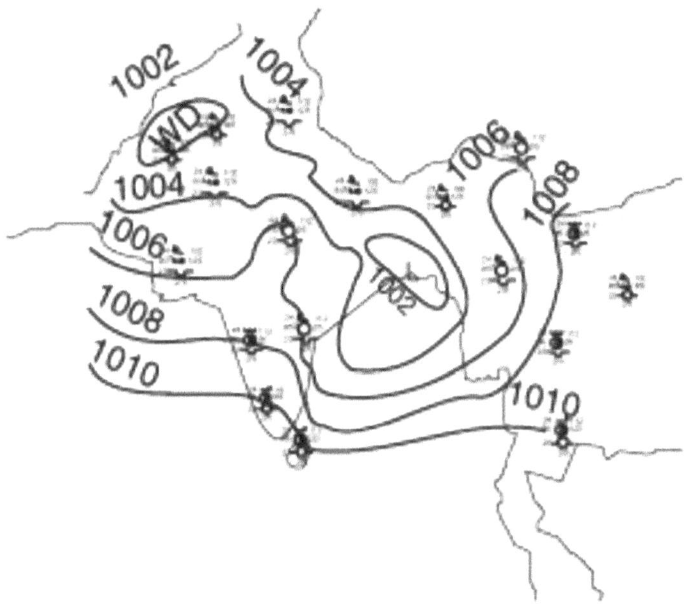

Fig. 83: Synoptic weather chart

Weather maps are created by plotting or tracing the values of relevant quantities such as sea level pressure, temperature and cloud cover onto a geographical map to find synoptic scale features. In the early days of meteorology, these pictures were all drawn by hands (Fig. 86) and now computers are used to compile the data to produce the weather maps. The most commonly used charts are synoptic charts and upper-air charts.

The surface charts are always studies jointly with a series of auxiliary charts depicting certain changes in weather elements. These charts show pressure, wind direction and dew point changes from an early observation. There are three kinds of charts to depict the atmospheric conditions at different levels.

i) *Surface chart*: The chart in which the geometric level is fixed and the distributions of pressure, temperature, humidity etc. are plotted is known as surface charts.

ii) *Pilot chart*: The charts on which upper-air wind's information are plotted is known as the pilot charts.

iii) *Constant pressure charts*: This is a chart in which isobaric level is fixed and plots the heights upto which one has to go vertically upwards from the sea level to reach that specified pressure level. The lines of equal values are known as contour lines. The heights are analyzed in the same manner as the isobaric analysis of the surface charts. They are normally prepared for 850, 700, 500, 300, 250, 200, 150, 100 and 50mb levels. The heights, as expresses on the constant pressure charts are not the geometric heights, but are known as geopotential heights and are measured in geopotential meters. The height and pressure table is shown in Table 14.

23.10. Geopotential meter

It is measure of energy possessed by a unit mass of air by virtue of its height above the sea level.

Table 14 : Geopotential height and pressure

Height in Kilometer	Pressure in millibar	Height in Kilometer	Pressure in millibar
0.0	1013	10.0	265
0.5	955	12.0	194
1	899	14.0	142
1.5	846	16.0	104
2.0	795	18.0	76
2.5	747	20.0	55
3.0	701	25.0	25
3.5	658	30.0	12
4.0	617	35.0	6
5.0	540	40.0	3
6.0	472	50.0	1
7.0	411		
8.0	356		
9.0	308		

The vast meteorological data collected simultaneously across the country or world are coded and plotted on the surface and upper-air charts to represent the current weather conditions, which is used to know the future weather.

23.11. Meteorological codes

Global scale meteorological observations within the shortest time are needed for forecasting of weather. Meteorological observations recorded in plain language are quite lengthy. Meteorological observations are transmitted globally, hence it should precise, universally readable irrespective of language, and therefore these data are converted into coded messages. These meteorological codes are specified by the World Meteorological Organisation (WMO). The weather data recorded at a station is transmission after coding these data over a station in a pre-specified place using weather codes (Fig. 84) to concise the data for its transmission.

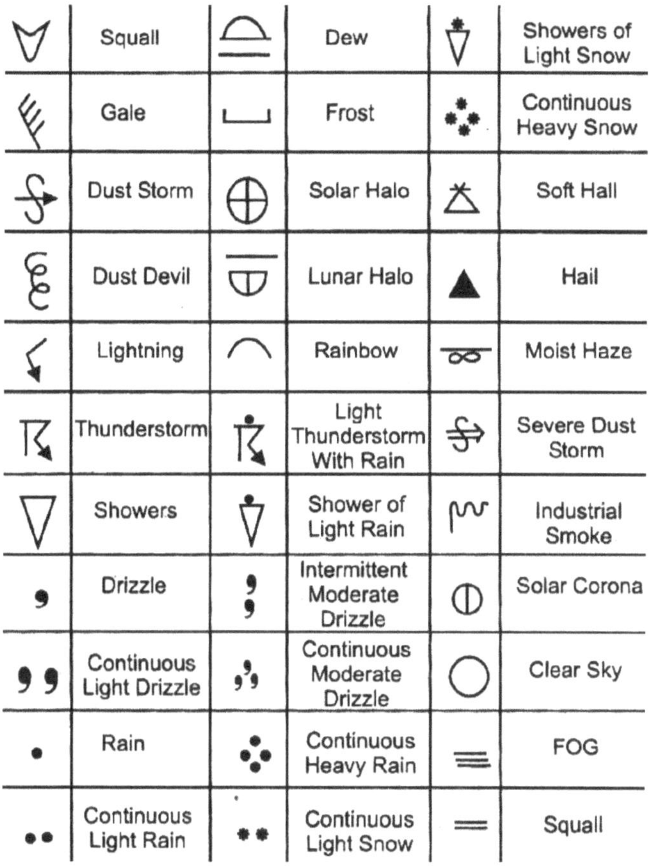

Fig. 84: Symbols for coding weather phenomena

When thousands of observations are to the processed, understood and inferences drawn before a forecast is issued, it would be convenient to plot the observations symbolically on a weather chart. A plot of message with the meaning of symbols is shown in figure 85. When all the data are plotted, the meteorologist analyses the charts by drawing lines of equal values of weather elements. Meteorologist is to delineate on the map the areas having identical meteorological parameters. This is achieved by drawing lines of isobars (equal pressure lines) figure 86 isotherm (equal temperature lines), isohyet (equal rainfall lines) isotach (equal wind speed lines). Lines connecting the mean direction of a wind at a given level on the weather charts are known as streamlines.

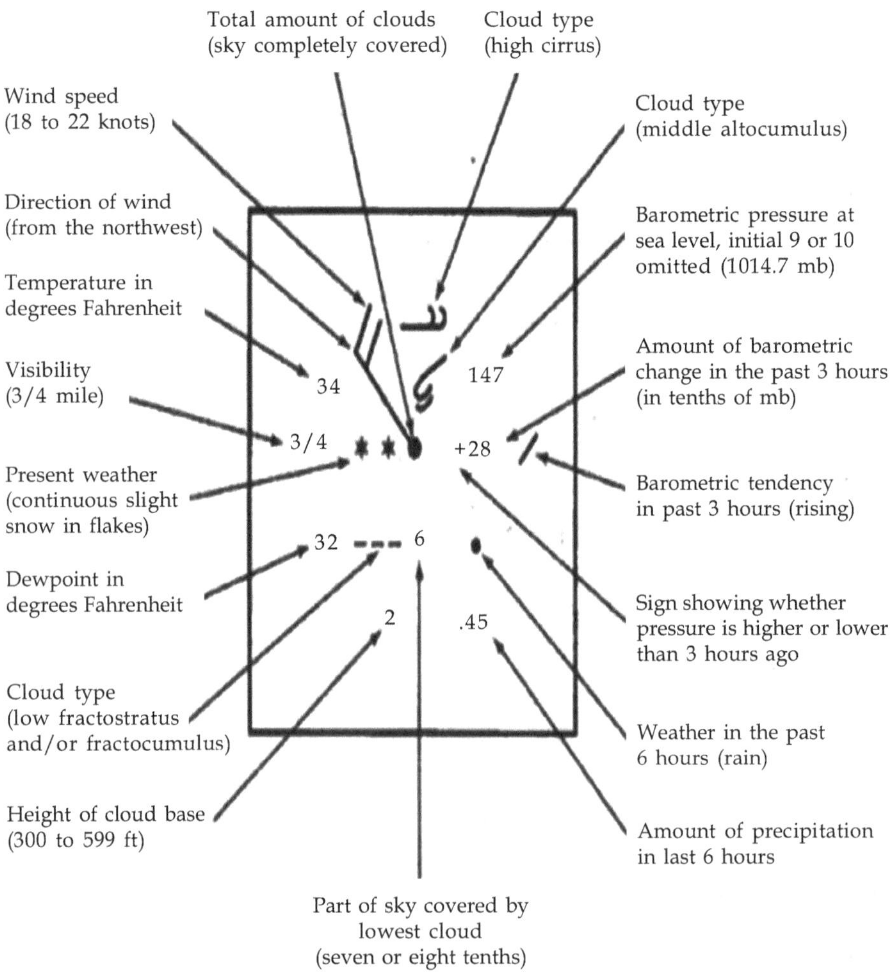

Fig. 85: Position of weather components plotted over a station using codes

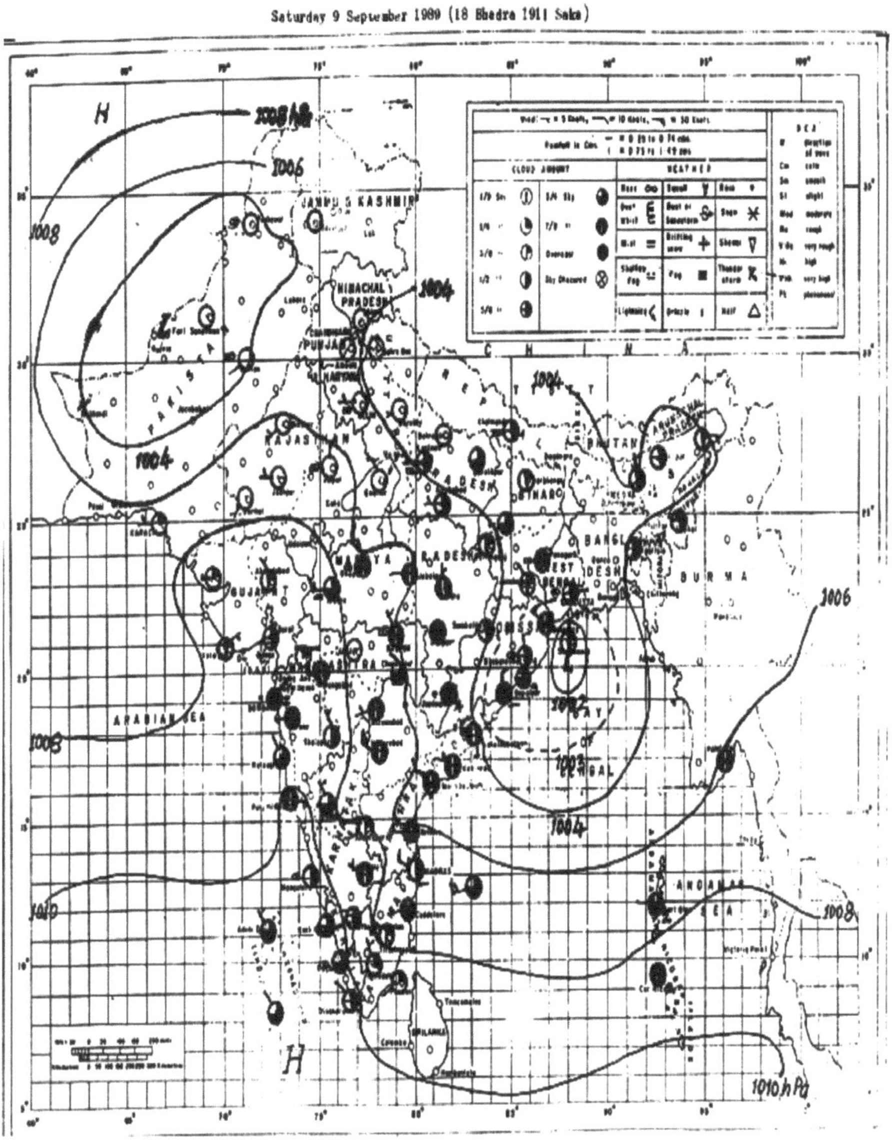

Fig. 86: Weather map depicting weather systems over India at 9 September, 1989
Source: IMD, New Delhi

Questions for Study

Q.1. What is the importance of upper-air data?

Q.2. What types of data are needed to represent the weather conditions at a place?

Q.3. Write the types of observatory used for weather data collection.

Q.4. What is a weather map?

Q.5. Why we need a geostationary satellite?

Q.6. Why weather code is needed in data transmission?

❑❑❑

Chapter 24

Agro-meteorological Data Management

Weather has profound influence on agriculture, which demands that the changes in weather conditions are regularly monitored. In qualitative terms, weather conditions are often expressed on the basis of final condition created by interaction of different components. For example, too hot, too cold, sunny or cloudy, humid, dry weather conditions refer to specific atmospheric conditions but are least useful in analyzing the impact of weather on agriculture.

The measurements taken at a weather station can include any number of atmospheric observables. The WMO has laid down specifications, procedures and practices in respect of various meteorological matters such as instruments, observations, telecommunications, processing of data, analysis etc.

24.1. Agro-meteorological Data Management

The agrometeorological data managements include data collection, reporting/transmission, storage and analysis which are utilized for its application in agricultural management, research and developments. These are described briefly in following paragraphs.

24.1.1. Data Collection

Quick collection of data from a vast network of surface and upper air observing stations, ships on high seas, aircraft reports, radar observations, forecasts and warnings and their selective distribution are vitally important and to meet national and international requirement. The National Meteorological Telecommunication Network (NMTN) is designed to collect observational data and to receive and distribute meteorological information on a national level. Basic data from the observatories are collected at either Meteorological Centers at the State Capitals or Regional Meteorological Centre by various telecommunication means. The stations whose data are exchanged on Global Telecommunication System are mostly connected by point to point links, telex or radio-telephony links. Observations from ships on high seas are collected at Regional Collecting Centers, Mumbai, Kolkata and Chennai through Coastal radio stations operated by the Department of Telecommunications.

24.1.2. Data recoding and reporting

Recorded weather data are entered into appropriate register and form for reporting to IMD and send to IMD Pune for its storage and archival at national data centre. The following type of register and forms are used for data reporting.

An observer record daily weather data from observatory are called raw data. These raw data are entered in a register called Pocket Register in appropriate columns. Correction factors on monthly basis, if any, may be also mentioned in this register. After corrections in the recorded data are entered in different register for reporting these registers are the permanent record. These register are given below:

- *Daily resister*: In this register, only the final observations, after applying required corrections in raw data are recorded. This should be done soon after the observations have been completed. It is a permanent record. In this register, the month and year are mentioned in the top of each page. First column is for dates and subsequent columns are meant for the observations pertaining to different observed weather elements. The observations which are recorded twice are mentioned as morning and evening observations. Daily register may be maintained in duplicate, one for transmission of data to the IMD Pune and another for permanent record.

- *Weekly register*: In the weekly register only weekly weather data is recorded. The daily data is first converted into weekly data. Daily data are converted into weekly total (by adding the values) and mean, by adding the values and dividing the number of days of respective weeks. The year is divided into 52 meteorological week starting from 1 week (from 1 January to 7 January). The 52nd and 9th weeks (in case of leap year) are of 8 days durations (Annexure-XVI). These 52 weeks are known as Standard Meteorological Week (SMW). The weekly data are entered into appropriate SMW columns in weekly register.
- *Monthly register*: The monthly data is computed from the daily data. The daily data, when added for a month and divided by the number of days becomes monthly data. But the monthly rainfall value is the total of daily rainfall values. After converting the daily data into monthly values, these data entered into appropriate columns in monthly register.
- *Crop Weather Station (CWS) form:* The form is same as daily register (DR) and data are copied from DR every day into CWS form. This form contains daily information of 4 to 5 meteorological weeks weather data. The morning and evening observations are recorded separately under the first hour (I) and second hour (II), respectively. Whenever the observation is taken only in morning, then it is entered only in first hour (I) column. Example - The maximum temperature is recorded twice a day but reported for only first hour (I). Week wise meteorological totals and means are computed and entered in appropriate places. The CWS form is two numbered, CWS 1 form and CWS 2 form. CWS 1 form is used for reporting of routine observatory observations (Annexure-XIV). CWS 2 form is used for reporting the microclimatic observations.

24.2. Data transmission

Modern meteorology depends upon near real time exchange of weather information across the entire globe. Telecommunication Division of India Meteorological Department provides the support function needed for meteorological data and processed products to the users, at both national and international levels. IMD maintains an extensive telecommunication network for speedy collection of meteorological information, both basic data and processed products over the globe. The main telecommunication hub of the IMD's telecommunication network is located at New Delhi. It is now termed as National Meteorological Telecommunication Centre (NMTC).

24.2.1. International Meteorological Telecommunication

Meteorological telecommunication of India Meteorological Department (IMD) consists of an integrated network of point-to-point circuits and multipoint circuits which inter connect meteorological centers within the country and the world for receiving data and relaying it selectively. It is mainly organized on a two level basis, The meteorological telecommunication network within the Global Telecommunication System (GTS) of World Weather Watch (WWW) program of World Meteorological Organization (WMO) and The National Meteorological Telecommunication Network (NMTN). IMD maintains an extensive telecommunication network with a central hub in Mausam Bhavan, New Delhi. The automated centre of RTH New Delhi is thus the Principal Meteorological Telecommunication Centres in South Asia and it collects observational data and feeds them on to the Global Telecommunication System for Global and Regional Exchange.

24.2.2. Digital Satellite Broadcasts

Satellite based data collection/distribution systems are integrated in the Meteorological Telecommunication Networks as an essential element of the global, regional and national level. Meteorological data products at 3 hourly interval (INSAT images, surface and upper air data, forecasted data, weather charts etc.) are exchanged over GTS.

A satellite data dissemination system (SADIS) is in operation at New Delhi to receive meteorological information from International Civil Aviation Organization (ICAO) centres which are routed to four international airports of India for national and international flight briefing and for providing data in GRIB/ BUFR format for wind/ temperature etc.

24.3. Primary handling of data

The weather station is part of a national network of meteorological/agro-meteorological data collection and storage system. The real time weather data is transmitted through land-line, radio or satellite and the data should be entered into a Database Management System (DBMS) and be made available to the users for further use. Some quality control (QC) of the data can be conducted locally where the data are being entered. Other quality controls such as spatial quality checks can be undertaken at the central database. It is important that all data, both raw and those processed for the long-term archive, should be backed up at every stage. The National Data Centre of IMD is located at IMD, Pune office for collection, storage and its distribution.

24.3.1. Quality control of data

It is necessary to weed out erroneous observations which might seriously degrade the quality of the data. All the data are checked for their consistency before entered it into computer or convert it into electronic media. Some of the quality control checks normally employed at Pune is listed below:

- Check of blank data, estimated mean value of the data is entered in blank data.
- Internal consistency checks on the data within the observation itself.
- Temporal consistency checks on the observations from the source.
- Climatological consistency checks.
- Checks for spatial consistency with neighboring observations.

24.4. Archival and availability of data

The national level weather data are recorded and communicated to national meteorological agency i.e., IMD. The storage and archival of voluminous weather data, which is continuously flowing from the recording stations to the national agency, is a challenging task. Before storing of these data, quality check is required. The weather data is first check for it consistency and quality, which is done at national meteorological data centre at IMD, Pune. The recorded weather data are than stored in electronic form for its easy archival and availability. Agromet data which is stored at IMD, Pune can be obtained after filled the prescribed format along with a request mentioning the type of required data, date and purpose of the use of data. The interested organisation, agency, institutes and other party may obtained the weather data on payment basis from surface meteorology, or National Data Centre, Shivaginagar, IMD, Pune (Annexure-XVII).

Questions for Study

Q.1. What type of correction is done before storing the weather data?

Q.2. Write a note on different type of registers used in data collection.

Q.3. Which agency in India is responsible for data collection and its storage?

Q.4. Write the data collection procedure from IMD Pune.

Q.5. What is raw weather data and can we store a raw data?

❑❑❑

Chapter 25

Agro-meteorological Data Analysis

The agro-meteorological data after the quality control is used to establish a relationship between the crop and weather or to understand their interactions. In agrometeorological data analysis, statistical tools are used. The frequent use of statistics terms and their computational formulas are given below for ready reference:

25.1. Mean

The word 'mean' denotes a representative of a whole set of observations "average" or "Arithmetic Mean (AM)".

AM: Sum of the values of variables divided by the number of variables in a set is known as mean or an average.

Given n observation $x_1, x_2, x_3 \ldots\ldots\ldots x_n$, their AM is given by

$$\bar{x} = \frac{\sum x}{n}$$

25.2. Range

The range is the simplest measure of variation to find. It is simply the highest value minus the lowest value.

Range = Maximum value - Minimum value

25.3. Standard deviation (S.D.)

It is square root of mean of the square of the deviation of each observation from their mean or S.D. of a set of observations is the square root of the arithmetic mean of squares of deviations from arithmetic mean.

$$\text{Standard deviation } (\sigma) = \sqrt{\frac{1}{n}\sum(xi-\bar{x})^2}$$

25.4. Variance

It is the square of standard deviation.

25.5. Co-efficient of variation (CV)

Coefficient of variation of a series of variate values is the ratio of the S.D. to the mean multiplied by 100. It measure of variation which is independent of the unit of treatment and is therefore useful for comparison between different series is known as coefficient of variation. It is relative measure of variation.

$$\text{C.V.} = \frac{\text{Standard deviation}(\sigma)}{\text{mean}(\bar{x})} \text{x}100$$

25.6. Correlation and Regression

25.6.1. Correlation coefficient (r)

It is a measure of intensity or degree of linear relationship between two variables. In a biovariate distribution, if the change in one variable affects a charge in the other variable, the variables are said to be correlated.

$$r = \frac{\Sigma(x-\bar{x})(y-\bar{y})}{\sqrt{\Sigma(x-\bar{x})^2\,\Sigma(y-\bar{y})^2}}$$

25.6.2. Regression

It is used to denote estimation or perdition of the average value of one variable for a specified value of the other variable. The estimation or prediction is done by means of equations and is known as regression equation.

The line of regression of y on x or x on y are given below:

$$y-\bar{y} = b(x-\bar{x})$$

or $y-\bar{y}=r\frac{\sigma y}{\sigma x}(x-\bar{x})$

Simmilarly $x-\bar{x}=r\frac{\sigma y}{\sigma x}(y-\bar{y})$

$byx=r\frac{\sigma y}{\sigma x}$ $bxy=r\frac{\sigma x}{\sigma y}$

byx and bxy are slopes of the lines of regressions.

To study the crop and weather relationship, it is required to convert the data into different form which are discussed below:

25.7. Standard Meteorological Week (SMW)

World meteorological organisation has divided a year into 52 standard (fixed date) weeks for meteorological purpose and these weeks are know as SMW. A standard meteorological week consists of 7 days. The first week starts from 1st January of a year and ends on 7th January, the 2nd SMW is starts from 8th January and ends on 14th January so on and so forth. The last SMW i.e 52 comprises of 8 days from 24th to 31st December of a year. In leap years, 9th SMW is also of 8 days from 26th February to 4th March. The SMW with date is given in Appendix -II.

25.8. Data analysis

The weather data are recorded on different time scales so it is necessary that they should be converted into different agro-meteorological scale for any meaningful analysis. Before any study we need the data on various scale as well as actual and normal values of the parameters. Knowledge of climatology of a place is very helpful for planning of various crop as well agricultural planning. Long-term (at least 30 years or more year data) averages of various weather parameters of a particular place/point are known as normals of that place and together constitute the climatology of that place.

25.8.1. Actual Values

It is a real value of weather elements which exist for any place and time. It is a value of particular weather element which has been recorded at a particular place at particular time.

25.8.2. Average values

It is mean or total of daily values over particular time scale of different weather elements of a place like weekly, decadal, and monthly values. Averages are calculated for all the weather elements to convert daily values into weekly, decadal, monthly except of rainfall. In case of rainfall only total is calculated to convert the daily values into weekly, decadal or monthly values. The daily value of different weather elements may of any time scale.

25.8.3. Normal Values

It is an average value of at least of 30 years or more of any weather elements. Therefore, normal of any weather elements is an average over 30 years or more periods.

25.9. Computation of data

25.9.1. Weekly data

Daily weather data is converted into weekly data by averaging it over a week or taking total (rainfall) for a week. The weekly data is converted according to SMW duration and used for further agrometeorological analysis. Weekly mean is calculated for all weather elements, except rainfall. Weekly rainfall value is a daily total value of a week; it is not a weekly mean value. However, if the observations are for only 3 days, no mean is calculated for that week.

25.9.2. Monthly data

Daily weather data of a month is converted into monthly data by taking average over the month or taking total (rainfall) of the month period.

25.9.3. Annual Data

Monthly data is converted into annual data by taking average for all months in a year or taking total (rainfall) of all months in a year.

25.10. Rainfall analysis

Rainfall is highly variable weather parameters and greatly influence crop grown and yield. Hence, rainfall analysis is mostly used in crop-weather study. The other derived parameters of rainfall are also used in agrometeorological study. Some of them is given below:

25.10.1. Rainy days

It is a day which has rainfall of 2.5 mm or more is called rainy day.

25.10.2. Assured rainfall

It is that rainfall which is assume to meet the evaporation or evapotraspiration (ET) demand of atmosphere or crop to particular place and time. It is calculated mostly on weekly or monthly basis. The mean weekly value of rainfall which is equal to the mean weekly ET of crop is computed on different probability, when weekly probability of rainfall is 75% or more then the corresponding weeks are categorized under assured rainfall period.

25.10.3. Weekly assured rainfall

Probabilistic analysis of weekly rainfall over India is very useful to the Plant breeders, Agronomists, Agriculture department, Planners, and the farmers for formulating policies regarding crop adaptation, plant breeding, drought management, irrigation planning and harnessing of rain water for maximizing agricultural production. Therefore, for efficient planning, research workers, farmers and planners stand to gain significantly by using quantified rainfall at different probability levels called Assured Rainfall. Incomplete Gamma distribution model is used for computing the assured rainfall amount at different decile levels.

Weekly rainfall of 50 mm with 70% or more probability of occurrence is called assured rainfall. Weekly assured rainfall of taluka level stations spread over various rainfall zones at 30, 40, 50, 60 and 70 percent probability levels has been prepared by IMD.

25.10.4. Dry and wet spell analysis

Climatological analysis in the form of probabilities is very useful in decision making for agricultural operations. Any information on the probability of weekly rainfall in any district of their concern based on pattern of actual rainfall during the preceding week help in framing need based advisory bulletin to the farmers. Based on this analysis, farmer are suggested to the whether the supplementary irrigation has to be provided or to refrain.

The interval between the onset of non effective rains and the commencement of the first wet spell is called the dry spell. Dry spell is duration of no or 1 mm of rainfall period. If the duration of dry spell exceeded a certain value, depending on the crop soil complex of the region, this spell is know as critical dry spell. The interval between the onset of effective rains and the commencement of the first dry spell is called the wet spell. A wet spell is defined as a rainy day with x mm of rainfall. X is the amount of rainfall which brings the top 50 cm soil layer to field capacity. The spell is generally represent a week.

25.10.5. Effective rainfall

Apart from soil, air and sunlight, crops need water to grow. The water need can be supplied to the crops either by rainfall or by irrigation or by a combination of rainfall and irrigation. Part of the rain water percolates below the root zone of the plants and part of the rain water flows away over the soil surface as run-off. This deep percolation water and run-off water cannot be used by the plants. The remaining part is stored in the root zone and can be used by the plants. This remaining part is the so-called effective rainfall. The factors which influence the effective rainfall are the climate, the soil texture, the soil structure and the depth of the root zone. Thus the effective rainfall is given by

Effective rainfall (ER) = A + B + C + D.

Where

A = water intercepted by living or dry vegetation

B = water lost by evaporation from the soil surface

C = the precipitation lost by evapo-transpiration during crop growth

D = water which contributes to leaching, percolation (D) or facilitates other cultural operations.

The effect of rainfall in lowering temperature and increasing humidity and its effect by advection on dry areas is not included.

25.11. Water requirement

Water requirement of a crop is meant the amount of water needed to raise it and this includes water to meet both consumptive and special needs, such as land preparation, land submergence, leaching and so on. The annual or seasonal effective rainfall as far as the water requirement of crops is concerned, should be interpreted as that portion of total annual or seasonal rainfall which is useful directly and/or indirectly for crop production at the site where it falls.

25.12. Crop weather calendar

It is calendar which provides information of different crop phases, extreme weather events and monthly normal of different weather elements at a place. This calendar is very helpful for the weather forecaster and also for issuing crop management advisories of a given place (Annexure-XV). The crop weather calendar has three parts *viz.*, crop husbandry, normal weather requirements and extreme weather information. The important crop phases like sowing, germination, tillering, elongation, flowering, grain formation and harvest are indicated under crop husbandry against the standard meteorological week. The components of crop weather calendars include: i) weather warning for the crops at various crop phenophases and ii) normal weather (standard week-wise) along with weather favorable for pests and diseases development.

25.13. Meteorological analysis

25.13.1. Onset of monsoon

Declaration of monsoon at place is based on rainfall, wind and radiation and detail of the criteria for onset of monsoon at a place is given below:

Rainfall

If after 10th May, 60% of the available stations of an area report rainfall of 2.5 mm or more for two consecutive days, the onset over that area be declared on the 2nd day.

Wind Field

Depth of westerlies should be maintained upto 600 hPa near by area. The zonal wind speed over the area should be of the order of 15 - 20 Kts. at 925 hPa.

Radiation (outgoing long wave radiation)

Derived outgoing long wave radiation (OLR) value should be below 200 wm^2 over the area. If at any palce the above conditions meet than onset of monsoon is declared over that area.

25.13.2. Onset of effective monsoon

The following criteria are proposed by Ashok Raj for determination of effective monsoon rainfall:

i) The first rain in a 7 days spell should not be less than average daily evaporation (e).

ii) The total rainfall during 7 days spell should not less than 5e+10mm

iii) At least four rainy days during 7 days spell.

25.14. Estimations of PET, reference crop evapotanspiration (ET_0), and AET

- *Potential evapotraspiration (PET):* The amount of water transpired in unit time by an extensive short green grass of uniform height completely shading the ground and never short of water. It is estimated by energy or water balance method.

PET can also be estimated by use of climatological variables in different expression of evapotranspiration and most common used methods are:

- Potential evapotranspiration by combination method such as penman (1948) or modified penman's methods.
- Water balance approach in providing the appropriate patterns of soil moisture depletion.
- Blaney-Criddle methods.
- Pan evaporation data with pan coefficients during different weather conditions.
- *Actual evapotarspiration (AET):* It the amount of water which is actual used by crop and soil for evaporation and transpiration purpose. It is determined by lysimeters.
- *Reference crop evapotranspiration (ET_0)* : It the rate of evapotranspiration from a hypothetical reference crop with an assumed height of 0.12 m, in a fixed surface resistance of 70 sec/m, and an elbedo of 0.23, actively growing, well watered and completely shading the ground. It is also know as reference evapotranspiration.
- *Crop coefficient (CC):* For computation of crop water requirement in a week, crop coefficient for a particular crop on different crop

stage is needed. The crop coefficient is calculated by PET and AET values.

CC=AET/PET

- Computation of potential evapotranspiration (PET) evaporation method.
    - Obtain the pan evaporation (E) from USWB-Class A open pan evaporimeter.
    - Calculate the weekly value of pan evaporation.
    - Pick up the appropriate pan coefficient (Kp) from Table given in Annexure-XVIII.
    - Multiply pan evaporation value and pan coefficient (i.e. by the formula ET_0 = KpXE), which gives PET value. Where Kp is the pan coefficient.
- Procedure for ET_0 calculation
    - Divide the total growing season of the crop into ten equal periods.
    - Compute mean pan evaporation (E) (in mm/day) by averaging daily values for the period.
    - Pick up the appropriate crop coefficient (CC) from Table given in annexure IV.
    - Calculate evapotranspiration of the crop (ET_0) for each of the ten periods by multiplying pan evaporation value and crop coefficient (i.e. by the formula ET_0 = CC x E).
    - Compute total water use of the crop during the whole growing season by adding all the ET_0 values of the ten periods.

25.15. Effective temperature

The effective temperature concept is used for relating the crop growth rate with air temperature. This concept assumed that relative growth rate declines linearly at supra-optimal temperature at the same rate as that at sub-optimal temperature. The effective temperature is given by

$$Te = To - |To - Ta|$$

and $Tb < Ta < Tc$

Where,

Te is effective temperature,

To is Optimum temperature

Ta is Mean temperature,

Tb is Base temperature

Tc is Ceiling temperature at which accumulation of thermal times ceases.

Questions for Study

Q.1. What is the importance of weather data in agriculture?

Q.2. What is a normal value and how it is calculated?

Q.3. How weekly and monthly values of rainfall is obtained from it daily values.

Q.4. How effective temperature is calculated?

Q.5. Compute reference evapotranspiration for the following data:

Mean pan evaporation	Crop coefficient
5.6 mm/day	0.94
7.1 mm/day	0.81
3.5 mm/day	1.16

❏❏❏

Chapter 26

Computation of Heat Indices

Heat units also called growing degree-days (GDD), effective heat units, or growing degree units, are a heuristic tool in phenology. GDD is a simple means of relating plant growth, development and maturity to air temperature. The concept is widely accepted as a basis for building phenology and population dynamic models. Degree day units are often used in Agrometeorology, Agronomy, Entomology and Pathology essentially to estimate or predict the lengths of the different phases of development in crop plants, insects, pathogens etc.

Unless stressed by other environmental factors like moisture, the development rate from emergence to maturity for many plants depends upon the daily air temperature. Because many developmental events of plants and insects depend on the accumulation of specific quantities of heat, it is possible to predict when these events should occur during a growing season regardless of differences in temperatures from year to year.

GDD can be used to assess the suitability of a region for production of a particular crop; estimate the growth-stages of crops, weeds or even life stages of insects; predict maturity and cutting dates of forage crops; predict best timing of fertilizer or pesticide application; estimate the heat stress on crops; plan spacing of planting dates to produce separate harvest dates. Crop specific indices that employ separate equations for the influence of the daily minimum (nighttime) and the maximum (day time) temperatures on growth are called crop heat units (CHUs).

26.1. Base Temperature (T_{base})

The base temperature is that temperature below which plant growth is zero. The base temperature, sometimes referred to minimum temperature, is an essential parameter for the determination of the beginning and the end of the growing season. It has therefore, an influence on the modeling of phenology and consequently of annual biomass increase. Generally base temperature value of 5.0 °C for winter crops and 10.0°C for summer season crops have been used.

26.2. Growing-degree day

Growing degree-days (GDD) is defined as the number of temperature degrees above a certain threshold base temperature, which varies among crop species. It is a measure of heat accumulation used by scientists to predict the date that a flower will bloom or a crop reach maturity. The GDD concept assumes a direct and linear relationship between growth and temperature. It assumes that the growth of a plant is dependent on the total amount of heat to which it is subjected during its lifecycle.

Growing degree days take aspects of local weather into account to predict the plants' pace toward maturity. GDD are calculated by taking the average of the daily maximum and minimum temperatures compared to a base temperature, T_{base}. Given an equation:

$$GDD = \frac{T_{max} + T_{min}}{2} - T_{base}$$

For example, a day with a high of 23 °C and a low of 12 °C (and a base of 10 °C) would contribute 7.5 GDDs.

$$\frac{23+12}{2} - 10 = 7.5$$

GDD are accumulated by adding each day's GDD contribution as the season progresses.

26.3. Accumulated degree days

Each developmental stage of an organism has its own total heat requirement. Development can be estimated by accumulating degree days between the high and low temperature thresholds throughout the season. Accumulation of degree days should be done regularly, especially when a control action decision is near.

26.4. Biofix

A biofix is the date to start accumulating degree days for a particular plant. It varies with the species and are usually based on specific biological events such as planting dates or first occurrence of a pest or the first date a weed seedling is observed or the date tree buds start to break. Degree days are reset to zero at the biofix date, no matter how many GDD is accumulated before that.

26.5. Calculation of different Heat Unit Systems

26.5.1. Growing Degree Days (GDD) or Heat Units

$$GDD = \sum_{a}^{b} (Tmax + Tmin)/2 - T_{base}$$

Where, $(T_{max} + T_{min})/2$ is the average daily temperature

T_{base} is the minimum threshold temperature for a crop

26.5.2. Photo-Thermal Units (PTU)

The product of degree day and maximum possible day length on any day is called photo thermal units.

PTU = GDD x Maximum Possible day length (°C days hours)

26.5.3. Helio-Thermal Units (HTU)

The product of degree day and the number of actual bright sunshine hours on any day is called helio-thermal unit.

HTU = GDD x Actual Sunshine Hours (°C days hours)

26.5.4. Heat Use Efficiency (HUE)

Heat use efficiency is used to compare the relative performance of different cultivars and treatments with respect to utilization of heat.

$$HUE = \frac{\text{Biomass acccumulated during phenophases } (g/m^2)}{\text{Heat units accumulated during the phenophase}}$$

26.5.5. Thermal use efficiency (TUE)

TUE = Biomass (kg/ha) ÷ GDD

26.5.6. Phenothermal index (PTI)

PTI = GDD ÷ Growth days

26.6. Merits of Growing Degree Days Concept

1. It is useful in forecasting of crop harvest date, yield and phenological stages.
2. The concept is also applied in growth and development of insect, pests and diseases.
3. It can be used to assess the suitability of a region for production of a particular crop and for the selection of a particular variety of a crop for specific region.
4. This concept can also be used to change or modify the microclimate in such away as to produce nearly optimum conditions at each point in the developmental cycle of an organism.

26.7. Demerits of Growing Degree Days Concept

1. Critical growth period of the crop has not been considered in the GDD concept.
2. Daily mean temperature is used instead of temperature range. The high temperature above optimum depresses crop growth, but the heat units increase.
3. Base temperature is used as a constant throughout the crop season. But practically base temperature changes with the change in crop stage.
4. The other weather parameters like rainfall, hails, wind, humidity also affects the crop growth and development which is not used in the GDD concept.
5. Soil fertility affects crop growth and development and is not accounted for in this concept.
6. The effects of topography, altitude, and latitude on crop growth are not taken into account.

Exercise: Calculate the Accumulated Growing Degree Days, Photo-Thermal Units and Helio-Thermal Units for germination of wheat crop using the following data.

Stage	Dates	Maximum temp (°C)	Minimum temp (°C)	Sunshine (hours)
Sowing date	2 Jan, 2006	13.6	8.8	1.0
	3 Jan, 2006	17.4	5.7	6.2
	4 Jan, 2006	16.0	5.5	4.1
	5 Jan, 2006	13.3	6.3	1.1
	6 Jan, 2006	16.0	4.2	3.5
	7 Jan, 2006	12.5	3.5	0.0
	8 Jan, 2006	17.9	5.5	5.8
	9 Jan, 2006	14.5	7.1	1.5
	10 Jan, 2006	14.1	5.9	3.5
Germination	11 Jan, 2006	14.8	6.3	4.5

Day length = 10 hrs, **Base temperature** (T_{base}) = 5.0 °C

Solution

1) Growing Degree Days (GDD)

$$GDD = \sum_{a}^{b} (T_{max} + T_{min})/2 - T_{base}$$

a = 2nd Jan., b = 11th Jan.

GDD = 6.2 + 6.6 + 5.8 + 4.8 + 5.1 + 3.0 + 6.7 + 5.8 + 5.0 + 5.6 = **54.6 days°C**

2) Helio-Thermal Units (HTU)

$$\textbf{HTU} = \sum_{a}^{b} (T_{mean} - T_{base}) \times SS$$

Where, a = 2nd Jan., b = 11th Jan.

$T_{mean} - T_{base}$ = Growing Degree Days (GDD),

SS = Sunshine hours

HTU = (6.2 x 1.0) + (6.6 x 6.2) + (5.8 x 4.1) + (4.8 x 1.1) + (5.1 x 3.5) + (3.0 x 0.0) + (6.7 x 5.8) + (5.8 x 1.5) + (5.0 x 3.5) + (5.6 x 4.5) = 6.2 + 40.9 + 23.8 + 5.3 + 17.9 + 38.9 + 8.7 + 17.5 + 25.2 = **184.4 days °C hrs**

3) Photo-Thermal Units (PTU)

$$PTU = \sum_{a}^{b}\left(T_{mean} - T_{base}\right) \times D$$

Where, A = 2nd Jan., b = 11th Jan.

$T_{mean} - T_{base}$ = Growing Degree Days (GDD)

D = Day length = 10 hours 36 minutes

= 10.6 hours

PTU = (6.2 x 10) + (6.6 x 10) + (5.8 x 10) + (4.8 x 10) + (5.1 x 10) + (3.0 x 10) +(6.7 x 10) + (5.8 x 10) + (5.0 x 10) + (5.6 x 10) = **546 days °C hours**

Questions for study

Q.1. What are degree days and they are used for?

Q.2. What is a biofix?

□□□

Glossary

Radiometric Terminology

Albedo	The portion of incoming radiation which is reflected by a surface
Azimuth angle	Angle in horizontal direction (0-360°)
Angle of incidence	Incident angle from zenith (vertical)
Cosine response	Detector response according to the cosine law
Diffuse solar irradiance	Solar radiation, scattered by water vapor, dust and other particles as it passes through the atmosphere
Direct solar irradiance	Radiation that has traveled a straight path from the sun
Global solar irradiance	Total irradiance falling on a horizontal surface
Irradiance	Radiant flux density (W/m^2)

Long-wave radiation	Radiation with wavelengths > 4 μm and < 100 μm
Pyrheliometer	Radiometer suitable to measure short-wave global radiation
Pyrgeometer	Radiometer suitable to measure downward long-wave radiation
Pyrheliometer	Radiometer suitable to measure direct irradiance
Short-wave radiation	Radiation with wavelengths > 280 nm and < 4 μm
Thermopile	Thermal detector made up of many thermocouple junctions
Zenith angle	Angle from zenith (0°, vertical)

Meteorological Terminology

Season

Winter season	January-February
Pre-Monsoon season	March-May
Southwest Monsoon season	June-September
Northeast Monsoon Season	October-December

Sky conditions

Clear sky	0 Okta
Mainly clear	1-2 Okta of sky covered
Partly cloudy	3-4 Okta of sky covered
Generally cloudy	5-7 Okta of sky covered
Cloudy	>7 Okta of sky covered

Weather Phenomena

Types	Description
One or two spells of rain	In a 24 hrs time, rainfall occurring with frequency of 1-2 spells.
A few spells of rain	In a 24 hrs time, rainfall occurring with a frequency of more than 2 spells but with well defined dry spells in between.
Intermittent rain	In a 24 hrs time, rainfall occurring with a frequency of more than that defined in "A few spells" but is discontinuous and without presenting the character of a shower.
Drizzle	Liquid precipitation in the form of water drops of very small size (by convection, with radius of water drops between about 100 and 500 mm).
Rain	Liquid precipitation in the form of water drops of radius between about 500 and 2500 mm).
Shower	Solid or liquid precipitation from a vertically developed cloud is designated a shower and is distinguished from the precipitation, intermittent or continuous, from layer clouds. Showers are often characterized by short duration and rapid fluctuations of intensity (by convention, with radius of water drops more than 2500 mm).
Hail	Solid precipitation in the form of balls or pieces of ice (hailstones) with diameters ranging from 5 to 50 mm or even more.
Thunderstorm	One or more sudden electrical discharges manifested by a flash of light (Lightning) and a sharp rumbling sound (thunder).
Dust storm	An ensemble of particles of dust or sand energetically lifted to great heights by a

	strong and turbulent wind. Surface visibility is reduced to low limits; the qualification for a synoptic report is visibility below 1000 m.
Mist	Mist is a phenomenon of small droplets suspended in air.
Dew	Condensation of water vapour on a surface whose temperature is reduced by radiational cooling to below the **DEW-POINT** of the air in contact with it
Fog	Fog is a phenomenon of small droplets suspended in air and the visibility is one kilometer or less.
Frost	Frost occurs when the temperature of the air in contact with the ground, or at thermometer-screen level, is below the freezing-point of water ('ground frost' or 'air frost', respectively). The term is also used of the icy deposits which may form on the ground and on objects in such temperature conditions (glaze, hoar-frost).
Haze	Haze is traditionally an atmospheric phenomenon where dust, smoke and other dry particles obscure the clarity of the sky.
Smog	Smoke and fog together reduce the visibility
Squally weather	Squally weather is meant to cover occasional or frequent squalls with rain or persistent type of strong gusty winds (mean wind speed not less than 20 knot) accompanied by rain. Such conditions are associated with low pressure systems or onset and strengthening of monsoon.
Visibility	The greatest distance that prominent object can be seen and identified by unaided, normal eyes.

Different Criteria for Weather Elements Categorization

Rainfall

Spatial Distribution of Rainfall

Distribution	No. of Places	Description
Isolated	One or two Places	<25% of stations gets rainfall
Scattered	At a few Places	(26-50)% of stations gets rainfall
Fairly Widespread	At many Places	(51-75)% of stations gets rainfall
Wide spread	At Most place	(76-100)% of stations gets rainfall
Dry	-	No station reported rainfall

Intensity of Rainfall

Descriptive Term used	Rainfall amount in mms
No Rain	0.0
Very light Rain	0.1 - 2.4
Light Rain	2.5 - 7.5
Moderate Rain	7.6 - 35.5
Rather Heavy	35.6 - 64.4
Heavy Rain	64.5 - 124.4
Very Heavy Rain	124.5 - 244.4
Extremely Heavy Rain	> 244.5
Exceptionally Heavy Rain	When the amount is a value near about the highest recorded rainfall at or near the station for the month or season. However, this term will be used only when the actual rainfall amount exceeds 12 cm.

Weekly/Seasonal Rainfall Distribution on Regional Scale

Excess — Percentage departure of realised rainfall from normal rainfall is + 20% or more.

Normal — Percentage departure of realised rainfall from normal rainfall is between – 19% to + 19%.

Deficient — Percentage departure of realised rainfall from normal rainfall is between – 20% to - 59%.

Scanty — Percentage departure of realised rainfall from normal rainfall is between - 60% to - 99%.

No Rain — Percentage departure of realised rainfall from normal rainfall is -100%

Rainfall Distribution on All India Scale

Normal — Percentage departure of realised rainfall is within ± 10% of the Long Period Average

Below Normal — Percentage departure of realised rainfall is < 10% of the Long Period Average

Above Normal — Percentage departure of realised rainfall is > 10% of the Long Period Average

All India Drought Year — When the rainfall deficiency is more than 10% and when 20 to 40% of the country is under drought conditions, then the year is termed as All India Drought Year

All India Severe Drought Year — When the rainfall deficiency is more than 10% and when the spatial coverage of drought is more than 40% it is called as All India Severe Drought Year.

Temperature

Description of Maximum temperature

(a) When the normal maximum temperature of a station is 40°C or below

Nomenclature	Past 24 hours change
Little change	-1°C to 1°C
Rise	2°C
Appreciable rise	3°C to 4°C
Marked rise	5°C to 6°C
Large rise	7°C or more

(b) When the normal maximum temperature of a station is more than 40°C

Nomenclature	Past 24 hours change
Little change	-1°C to 1°C
Rise	2°C
Marked rise	3°C to 4°C
Large rise	5°C or more

Description of Minimum temperature

(a) When the normal minimum temperature of a station is 10°C or more

Nomenclature	Past 24 hours change
Little change	1°C to -1°C
Fall	-2°C
Appreciable fall	-3°C to -4°C
Marked fall	-5°C to -6°C
Large fall	-7°C or less

(b) When the normal minimum temperature of a station is more than 10°C

Nomenclature	*Past 24 hours change*
Little change	1°C to -1°C
fall	-2°C
Marked fall	-3°C to -4°C
Large fall	-5°C or more

Criteria for Heat Wave

Heat wave need not be considered till maximum temperature of a station reaches at least 40°C for Plains and at least 30°C for Hilly regions.

a) When normal maximum temperature of a station is less than or equal to 40°C

Heat Wave Departure from normal is 5°C to 6°C

Severe Heat Wave Departure from normal is 7°C or more

b) When normal maximum temperature of a station is more than 40°C

Heat Wave Departure from normal is 4°C to 5°C

Severe Heat Wave Departure from normal is 6°C or more

c) When actual maximum temperature remains 45°C or more irrespective of normal maximum temperature, heat wave should be declared.

Hot Day

Whenever, the maximum temperature remains 40°C or more and minimum temperature is 5°C or more above normal, it may be defined as Hot Day, provided it is not satisfying the heat wave criteria given above.

Criteria for describing Hot Day for coastal stations

When maximum temperature departure is 5°C or more from normal, Hot Day may be described irrespective of the threshold value of 40°C. If the threshold value of 40°C is reached Heat Wave may be declared.

When a station satisfies both the Heat Wave and Hot Day criteria, then Heat Wave should be given higher priority and be declared.

Criteria for cold wave

Wind chill factor plays an important role and brings down the actual minimum temperature depending upon the wind speed. The actual minimum temperature of a station should be reduced to "wind chill effective minimum temperature (WCTn)" based on wind chill factor.

If WCTn is 10°C or less, then only the conditions for cold wave should be considered.

a) When normal minimum temperature is equal to 10°C or more.
   Cold Wave Departure from normal is -5°C to -6°C.
   Severe Cold Wave Departure from normal is – 7°C or less

b) When normal minimum temperature is less than 10°C.
   Cold Wave Departure from normal is -4°C to -5°C.
   Severe Cold Wave Departure from normal is -6°C or less.

❑❑❑

Annexures

Annexure-I : Conversion of Temperature in different scales

Celsius	Fahrenheit	Kelvin	Reaumer
1°C	1.8 °C+32	1°C+273	0.8 °C
0.5556°F-17.7779	1°F	0.5556°F+255.221	0.4445 °F-14.224
1°K-273	1.8 °K-459.4	1°K	0.8 °K-218.4
1.25 °R	2.25 °R+32	1.25 °R+273	1°R

Annexure-II : Conversion of units of pressure

Atmosphere	dynes/cm^2	Pascal	mbar	mm of Hg
1	1013×10^3	1103×10^2	1013	760
0.987×10^{-6}	1	0. 1	0.001	0.75×10^{-3}
0.987×10^{-5}	10	1	0.01	0.75×10^{-2}
0.987×10^{-3}	1000	100	1	0.75
0.13157×10^{-2}	1.333×10^3	1.333×10^2	1.333	1

Annexure-III : Inter conversion of units of wind

cm/s	m/s	km/hr	mile/hr	knots
1	0.0	0.036	0.0223	0.01943
100	1	3.6	2.2347	1.943
27.78	0.2778	1	0.6214	0.5396
44.77	0.4477	1.6109	1	0.8696
51.48	0.5148	1.8533	1.1516	1

Annexure-IV : Crop coefficient of important *kharif* and *rabi* crops

Stage of crop growth (in % of crop growing season)	Wheat	Sorghum	Maize	Cotton	Rice	Groundnut
1-10	0.18	0.44	0.43	0.22	1.02	0.31
10-20	0.34	0.48	0.55	0.24	1.05	0.36
20-30	0.57	0.53	0.74	0.40	1.09	0.50
30- 40	0.80	0.60	0.95	0.78	1.13	0.69
40-50	0.90	0.71	1.07	1.02	1.18	0.84
50-60	0.90	0.91	1.10	1.08	1.21	0.93
60-70	0.86	1.06	1.09	1.04	1.22	0.95
70-80	0.80	1.08	1.04	0.93	1.19	0.90
80-90	0.71	0.98	0.95	0.73	1.10	0.79
90-100	0.58	0.81	0.80	0.51	0.95	0.64

(Adapted from the report by Water Management Division, Department of Agriculture and Irrigation, Government of India, 1971).

Annexure-V : Albedo values of natural surfaces for total short wave radiation

Surface	Albedo (%)
Bare field	12-25
Fresh snow	75-95
Water	3-10
Sand	35
Granite	15
Dark wet clay	2-8
Dark dry clay	16
Sand wet	18
Sand dry	18
Wet ploughed fields	5-24
Stubble fields	15-17
Green grass	16-27
Dried grass	16-19
Grain crops	10-15
Desert mid-day	15
Forests	5-20

Source: Ghildyal and Tripathi, 1987

Annexure-VI : Conversion formulae

- Conversion of temperature values in 0°C to °F or °F to °C by the formula:

$$\frac{C}{5} \quad \frac{F-32}{9}$$

- Conversion of temperature values in °C to °K

K = C + 273
C = K-273

- Conversion of temperature values for °C to °R

$$\frac{C}{5} \quad \frac{R}{4}$$

Annexure: VII- Units and their conversion into other units

Pressure (Pa, bar, mm Hg)

Pascal, Pa	=	(2.09×10^{-2}) Pounds per square foot, 1 b/ft^{-2}
Pascal, Pa	=	(1.45×10^{-4}) Pounds per square inch, 1.lb inch^{-2}
Pascal, Pa	=	1N.m^{-2}; 1 bar = 10^6 dynes cm^{-2}
Pascal, Pa	=	0.01 mb; 1 bar = 10^3 millibars (mb)
100 Pascal	=	1 hecto-Pascal (hPa); 1mb = 100 Pascals (Pa)
1000 Pascal	=	1 Kilo Pascal (KPa); 1 mb = 1000 dynes cm^{-2}
Mega Pascal, Mpa	=	10^6 Pascal = 10 bar
1kPa	=	10 mb; 1mb = 0.750 mm Hg
1 mm	=	1.33 mb; 1 mb = 1 hecto Pascal, hPa
1 mm	=	133.32 Pa
1 Standard Atmosphere	=	30 inch = 1013 mb = 101.32 kPa = 76 cm or 760
mm of Hg pressure	=	14.7 lb.inch^{-2} = 1.04×10^6 dynes cm^{-2}

Wind velocity (kt, ms^{-1}, mPh, KPh)

1 Knot, Kt = 1 Nautical mile
1 Knot, Kt = 1.15 mile per hour (mph); 1 ms^{-1} = 3.60 kmh^{-1}
1 knot, Kt = 0.515 ms^{-1} 1ms^{-1} = 2.24 mile h^{-1}
1 Knot, Kt = 1.853 Km per hour (KPh); 1 ms^{-1} = 1.94 Knots 1 mPh = 1.609 Km h^{-1};
1mph = 0.8684 Knots
1 mph = 0.447 ms^{-1}

Humidity

RH expressed in %, AH in mb, mm, hPa, kPa
SH in g of the Kg^{-1}, etc. Vapour Pressure or SVP in 1kPa = 10mb,
Water Potential, 1MPa = 10 bars, slope of SVP with temp. 1kPa K^{-1} = 10mb K^{-1}

Power (Joule, Calorie, Watt)

Joule, 1J = 10^7 erg; Watt, 1W = $1JS^{-1}$
1J = 0.74 foot-pound; 1W = 0.239 cal S-1
1J = 0.24 Calori, Cal; 1W = 0.0569 BThv min $^{-1}$
1W = 2.39 x 10^{-5}1y $m^{-2}s^{-1}$;1W = 0.00134
Electrical Horse Power 1Wm^{-2} = 1.4 x 10^{-3} $Cal.cm^{-2}$, min^{-1}

Energy (Joule, Calorie, Watt)

Calorie, 1 Cal = 4.187 J; Δ gram - Calorie = 4.187 x 10^7 erg
1 Cal = 4.187 x 10^7 erg; $1J.m^{-2}$ = 2.38 x 10^5 $Cal.cm^{-2}$
1 Cal = 1.16 x 10^6 KWh^{-1}; $1J.m^{-2}$ = 2.38 x 10^5 Ly (Langley)
1 Cal = 3.97 x 10^{-3} BThU; 1Cal = 1 longley cm^{-2}

Radiation (Joule, Calorie, Watt μEi)

A Radiation Flux Density: $1J.m^{-2}.S^{-1} = 1W.m^{-2}$,
1 Cal = 4.184 J, $1Cal.cm^{-2}$ = 1 Langley (Ly)
1 $Cal.cm^{-2}.min^{-1}$ = 697.8wm^{-2}, 1W = $1J.s^{-1}$
1 Photon = 1 quantum of light, 1 Einstein (E) = 6.023x10^{23} quanta,
1ìE = 6.023 x 10^{17} quanta, 1 $Cal.cm^{-2}$ = 1/59mm (water evaporation)
1 mm h^{-1} (water evaporation) = 59 $Ca1.Cm^{-2}$ h^{-1}
= 68.9 $milliwatt.cm^{-2}$

Light intensity (Lux, Foot-candle)

1 kLux = 1000 Lux; 1 foot Candle (F.C.) = 10.76 Lux
1 FC = 0.044 wm^2 of PAR Light, 1wm^{-2} = 4.7 μEm^2S^{-1}
sunlight 1 Lux = 1 Lumen m^{-2} = 0.00041 wm^{-2} of PAR

Quality (Wavelength): μm, mm, Å

1 μm = 10^{-6}m, 1 nanometer (nm) = 10^{-9}m,
1Angstom (N) = 10^{10}m 1wm^{-2} = 4.6 μmol m^{-2} s^{-1} = 0.25 kLux of PAR Daylight.

Photosynthesis

$mg.m^{-2}.s^{-1}$ of CO_2 = 0.036$g.dm^{-2}.s^{-1}$.

Transpiration

$mg.m^{-2}.s^{-1}$ of H_2O = 5.5 X 10^{-3} $\mu mol.cm^{-2}s^{-1}$

Annexure-VIII : Useful Physical Constants

Name of the Constant	Abbreviation	Value and Units
Universal Gas Constant	R*	8.3143×10^{3} J K^{-1} $molecule^{-1}$
Boltzmann's Constant	k	1.381×10^{-23} J K^{-1} $molecule^{-1}$
Avogadro's Number	N_A	6.022×10^{23} mol^{-1}
Stefan-Boltzmann Constant	s	5.6696×10^{-8} W m^{-2} K^{-4}
Planck's Constant	h	6.6262×10^{-34} J s
Velocity of Light	c	2.998×10^{8} m s^{-1}
Permittivity of Vacuum	e_o	8.85×10^{-12} C^2 N^{-1} m^{-2}
Average Radius of the Earth	R_E	6.37×10^{6} m
Mean Gravity at Earth's Surface	g_o, g	9.81 m s^{-2}
Angular Velocity of Rotation of the Earth	W	7.292×10^{-5} rad s^{-1}
Average Distance from the Center of the Sun to the Center of the Earth	d, d_o	1.50×10^{11} m
Solar Irradiance on a Perpendicular Plane at the mean Earth-Sun Distance	S_o	1.367×10^{3} W m^{-2}
Average Molecular Weight of Dry Air	M_d	28.97
Gas Constant of Dry Air	R_d	287 J K^{-1} kg^{-2}
Density of Dry Air at 0°C and 1000mb	r_d	1.275 kg m^{-3}
Specific Heat of Dry Air at Constant Pressure	c_{pd}	1004 J K^{-1} kg^{-1}
Specific Heat of Dry Air at Constant Volume	c_{vd}	717 J K^{-1} kg^{-1}
Thermal Conductivity of Dry Air at 0°C	k	2.40^{-2} J m^{-1} s^{-1} K^{-1}
Molecular Weight of Water	M_w	18.016
Gas Constant for Water Vapor	R_v	461 J K^{-1} kg^{-1}
Density of Liquid Water at 0°C	r_w	1.00×10^{3} kg m^{-3}
Density of Ice at 0°	r_i	0.917×10^{3} kg m^{-3}
Specific Heat of Water Vapor at Constant Pressure	c_{pv}	1952 J K^{-1} kg^{-1}
Specific Heat of Water Vapor at Constant Volume	c_{vv}	1463 J K^{-1} kg^{-1}
Specific Heat of Water Vapor at 0°C	c_w	4218 J K^{-1} kg^{-1}
Specific Heat of Ice at 0°C	c_i	2106 J K^{-1} kg^{-1}
Enthalpy (Latent Heat) of Vaporization of Water at 0°C	l_v	2.50×10^{6} J kg^{-1}
Enthalpy (Latent Heat) of Vaporization of Water at 100°C	l_v	2.25×10^{6} J kg^{-1}
Enthalpy (Latent Heat) of Fusion of Water at 0°C	l_f	3.34×10^{5} J kg^{-1}
Enthalpy (Latent Heat) of Sublimation of Water at 0°C	l_s	2.83×10^{6} J kg^{-1}

Annexure-IX : Beaufort Scale for wind speed estimation

Beaufort Scale	Description	Wind Speed in Knots	Specifications for estimating over land
0	Calm	<1	Calm : smoke rises vertically
1	Light air	1-3	Direction shown by smoke-drift but not by wind vane.
2	Light breeze	4-6	Wind felt in face; leaves rustle; ordinary vanes moved by wind
3	Gentle breeze	7-10	Leaves/small twigs in constant motion; wind extends light flag
4	Moderate breeze	11-16	Raises dust and loose paper ; small branches are moved
5	Fresh breeze	17-21	Small trees begin to sway ; created wavelets from on inlands waters
6	Strong breeze	22-27	Large branches in motions; whistling heard in telegraph wires; umbrellas used with difficulty
7	Near gale	28-33	Whole trees move ; inconvenience in walking against wind
8	Gale	34-40	Breaks twigs off trees ; generally impedes progress
9	Strong gale	41-47	Slight structural damage occurs(chimney-pots and slates removed)
10	Storm	48-55	Seldom experienced inland; trees uprooted ; considerable structural damage occurs.
11	Violent storm	56-63	Very rarely experienced ; accompanied by widespread damage
12	Hurricane	>63	Hurricane conditions.

Annexure-X : Mean monthly duration of Maximum Possible Sunshine Hours (N) for different months and latitudes

Northern Lats	Jan	Feb	Mar	Apr	May	June	July	Aug	Sept	Oct	Nov	Dec
Southern Lats	**July**	**Aug**	**Sept**	**Oct**	**Nov**	**Dec**	**Jan**	**Feb**	**Mar**	**Apr**	**May**	**June**
50°	8.5	10.1	11.8	13.8	15.4	16.3	15.9	14.5	12.7	10.8	9.1	8.1
48°	8.8	10.2	11.8	13.6	15.2	16.0	15.6	14.3	12.6	10.9	9.3	8.3
46°	9.1	10.4	11.9	13.5	14.9	15.7	15.4	14.2	12.6	10.9	9.5	8.7
44°	9.3	10.5	11.9	13.4	14.7	15.4	15.2	14.0	12.6	11.0	9.7	8.9
42°	9.4	10.6	11.9	13.4	14.6	15.2	14.9	13.9	12.6	11.1	9.8	9.1
40°	9.6	10.7	11.9	13.3	14.4	15.0	14.7	13.7	12.5	11.2	10.0	9.3
35°	10.1	11.0	11.9	13.1	14.0	14.5	14.3	13.5	12.4	11.3	10.3	9.8
30°	10.4	11.1	12.0	12.9	13.6	14.0	13.9	13.2	12.4	11.5	10.6	10.2
25°	10.7	11.3	12.0	12.7	13.3	13.7	13.5	13.0	12.3	11.6	10.9	10.6
20°	11.0	11.5	12.0	12.6	13.1	13.3	13.2	12.8	12.3	11.7	11.2	10.9
15°	11.3	11.6	12.0	12.5	12.8	13.0	12.9	12.6	12.2	11.8	11.4	11.2
10°	11.6	11.8	12.0	12.3	12.6	12.7	12.6	12.4	12.1	11.8	11.6	11.5
5°	11.8	11.9	12.0	12.2	12.3	12.4	12.3	12.3	12.1	12.0	11.9	11.8
0°	12.0	12.0	12.0	12.0	12.0	12.0	12.0	12.0	12.0	12.0	12.0	12.0

Annexure-XI : Extra Terrestrial radiation (Ra) expressed in equivalent evaporation in mm/day

	Northern hemisphere											
Latitude	**Jan**	**Feb**	**Mar**	**Apr**	**May**	**June**	**July**	**Aug**	**Sept**	**Oct**	**Nov**	**Dec**
50°	3.8	6.1	9.4	12.7	15.8	17.1	16.4	14.1	10.9	7.4	4.5	3.2
48°	4.3	6.6	9.8	13.0	15.9	17.2	16.5	14.3	11.2	7.8	5.0	3.7
46°	4.9	7.1	10.2	13.3	16.0	17.2	16.6	14.5	11.5	8.3	5.5	4.3
44°	5.3	7.6	10.6	13.7	16.1	17.2	16.6	14.7	11.9	8.7	6.0	4.7
42°	5.9	8.1	11.0	14.0	16.2	17.3	16.7	15.0	12.2	9.1	6.5	5.2
40°	6.4	8.6	11.4	14.3	16.4	17.3	16.7	15.2	12.5	9.6	7.0	5.7
38	6.9	9.0	11.8	14.5	16.4	17.2	16.7	15.3	12.8	10.0	7.5	6.1
36	7.4	9.4	12.1	14.7	16.4	17.2	16.7	15.4	13.1	10.6	8.0	6.6
34	7.9	9.8	12.4	14.8	16.5	17.1	16.8	15.5	13.4	10.8	8.5	7.2
32	8.3	10.2	12.8	15.0	16.5	17.0	16.8	15.6	13.6	11.2	9.0	7.8
30	8.8	10.7	13.1	15.2	16.5	17.0	16.8	15.7	13.9	11.6	9.5	8.3
28	9.3	11.1	13.4	15.3	16.5	16.8	16.7	15.7	14.1	12.0	9.9	8.8
26	9.8	11.5	13.7	15.3	16.4	16.7	16.6	15.7	14.3	12.3	10.3	9.3
24	10.2	11.9	13.9	15.4	16.4	16.6	16.5	15.8	14.5	12.6	10.7	9.7
22°	10.7	12.3	14.2	15.5	16.4	16.4	16.4	15.8	14.6	13.0	11.1	10.2
20	11.5	12.7	14.4	15.6	16.3	16.4	16.3	15.9	14.8	13.3	11.6	10.7

18	11.6	13.0	14.6	15.6	16.3	16.1	16.1	15.8	14.9	13.6	12.0	11.1
16	12.0	13.3	14.7	15.6	16.1	15.9	15.9	15.7	15.0	13.9	12.4	11.6
14	12.4	13.6	14.9	15.7	15.9	15.7	15.7	15.7	15.1	14.1	12.8	12.0
12	12.8	13.9	15.1	15.7	15.7	15.5	15.5	15.6	15.2	14.4	13.3	12.5
10	13.2	14.2	15.3	15.7	15.5	15.3	15.3	15.5	15.3	14.7	13.6	12.9
8	13.6	14.5	15.3	15.6	15.3	15.0	15.1	15.4	15.3	14.8	13.9	13.3
6	13.9	14.8	15.4	15.4	15.1	14.7	14.9	15.2	15.3	15.0	14.2	13.7
4	14.3	15.0	15.5	15.5	14.9	14.4	14.6	15.1	15.3	15.1	14.5	14.1
2	14.7	15.3	15.6	15.3	14.9	14.2	14.3	14.9	15.3	15.3	14.8	14.4
0	15.0	15.5	15.7	15.3	14.4	13.9	14.1	14.8	15.3	15.4	15.1	14.8

Annexure-XII : Monthly values of maximum possible sunshine hours at different station of India

Station	Jan	Feb	Mar	Apr	May	June	July	Aug	Sept	Oct	Nov	Dec
Bhopal	9.94	10.51	11.16	11.81	12.34	12.61	12.50	12.06	11.46	10.78	10.14	9.79
Gwalior	9.73	10.39	11.12	11.88	12.49	12.79	12.67	12.16	11.47	10.69	9.96	9.56
Indore	9.98	10.54	11.16	11.80	12.32	12.57	12.47	12.04	11.45	10.79	10.18	9.84
Jabalpur	9.95	10.52	11.16	11.51	12.34	12.60	12.49	12.06	11.46	10.78	10.15	9.80
Kota	9.81	10.43	11.13	11.85	12.44	12.72	12.61	12.13	11.46	10.72	10.03	9.65
Lucknow	9.70	10.36	11.11	11.89	12.51	12.82	12.70	12.18	11.47	10.67	9.93	9.52
Nagpur	10.09	10.60	11.18	11.77	12.24	12.47	12.38	11.99	11.45	10.84	10.27	9.95
New Delhi	9.56	10.28	11.09	11.93	12.61	12.94	12.81	12.25	11.47	10.61	9.81	9.37
Patna	9.78	10.41	11.13	11.86	12.46	12.75	12.63	12.14	11.46	10.71	10.00	9.61
Pune	10.25	10.70	11.20	11.71	12.12	12.32	12.24	11.91	11.44	10.91	10.41	10.14
Kolkatta	9.99	10.54	11.16	11.80	12.31	12.57	12.46	12.04	11.45	10.80	10.18	9.84
Chennai	10.58	10.90	11.25	11.60	11.87	12.00	11.95	11.73	11.41	11.05	10.69	10.50
Bhubanesh-war	10.14	10.63	11.19	11.75	12.20	12.42	12.33	11.96	11.44	10.86	10.32	10.02
Srinagar	9.10	10.00	11.01	12.05	12.90	13.33	13.15	12.45	11.48	10.42	9.41	8.86
Jaipur	9.69	10.36	11.11	11.89	12.52	12.83	12.70	12.18	11.47	10.67	9.93	9.52
Dehradun	9.42	10.19	11.07	11.97	12.70	13.06	12.91	12.31	11.48	10.56	9.69	9.21
Agartala	9.91	10.49	11.15	11.82	12.37	12.64	12.53	12.08	11.46	10.76	10.11	9.75
Guwahati	9.74	10.39	11.12	11.88	12.49	12.79	12.66	12.16	11.47	10.69	9.97	9.57
Hyderabad	10.32	10.74	11.21	11.69	12.07	12.26	12.18	11.87	11.43	10.94	10.47	10.21
Port Blair	10.66	10.94	11.26	11.57	11.81	11.93	11.88	11.69	11.40	11.08	10.76	10.58
Ranchi	9.93	10.51	11.15	11.82	12.33	12.61	12.51	12.07	11.46	10.77	10.13	9.78

Annexure-XIII : Important Agriculture Meteorology/ Meteorology Journals

International Journals

1. Agricultural and Forest Meteorology
2. Agrometeorological Bulletin
3. Meteorological Application
4. Journal of Agricultural Meteorology

5. Journal of Applied Meteorology
6. Agronomy Journal
7. International Journal of Biometeorology
8. Journal of Agricultural Meteorology
9. Italian Journal of Agrometeorology
10. Agriculture, ecosystems & Environment
11. Journal of Climate
12. Monthly Weather Review
13. New Journal of Water and Climate
14. Weather, Climate, and Society
15. Journal of Applied Climatology

National Journals

16. Journal of Agrometeorology
17. Mausam
18. Vayu Mandal

Annexure-XIV : Meteorological / Weather Sheet

C.W.S.-1 **MET-1 (Agro)-1**

Meteorological Observatory at.........................

Year.............
I. 0700 hrs. local mean time **Page-1**
II. 1400 hrs. local mean time

Latitude Longitude height above mean sea level Meters Hours of observation hrs. IST hrs. IST

6,7 8, 9 12 13 14, 15 16 17, 18-20 21-23 24-26 27-29 30-32 33-35 36-38 39-41 42-44 45-47 48-50 51-53 54-56 57-58 59-60 61 62 63 64 65 66 67 68 69 70 71 72 73 74 75 77 78 79 80

			Temperature °C (Corrected)							**Soil temperature (Corrected)**						**Humidity**				**Wind**			**Sunshine**	**Rain**	**Weather**		**Evaporation**
																					Direction at						
Period No.	Week No.	Date	Dry	Wet	Dry	Wet	Max.	Min.	Gras Min.	5/7½	10/15	20/30	5/7½	10/15	20/30	Vapour pressure in mm of mercury		Humidity (%)		Mean velocity in past 24 hrs. ending at 0700 hrs Km. p.hr.	0700 hrs.	1400 hrs.	No. Of hrs. Of sunshine during previous day	In past 24 hrs. ending at 08-30 hrs. (mm)	0700 hrs.	1400 hrs.	In past 24 hrs. ending at 08-30 hrs. IST USA (m.m.)
			I	I	II	II	I	II	I	I	I	I	II	II	II	I	II	I	II	I	I	II	I	I	I	II	I
(1)	(2)	(3)	(4)	(5)	(6)	(7)	(8)	(9)	(10)	(11)	(12)	(13)	(14)	(15)	(16)	(17)	(18)	(19)	(20)	(21)	(22)	(23)	(24)	(25)	(26)	(27)	(28)

contd...

Total Mean																				
Total																				
Mean																				
Total																				
Mean																				

Page -2

WEATHER DIARY

THE INTENSITY OF PHENOMENA SHOULD BE RECORDED AS INDICATED BELOW

	Light	**Moderate**	**Heavy**	Thunder storm	Light peals of thunder and lighting at long intervals	Frequent flashes of lighting & thunder with moderate rainwind 30-70 km/hr.	Continuous thunder & lighting with heavy rainwind more than 70 km/hr.
Rain	Less than 13 mm/hr	13 mm to 38 mm/hr.	38 mm and more				
Snow				Hail storm			
Dew	Small droplets seen on exposed objects	Big droplets seed on exposed objects	Big droplets collect and fall from exposed objects		**Thin**	**Moderate**	**Thick**
				Fog	Visibility less than 1 km.	Visibility less than 1/2 km.	Visibility less than 200 mts.
	Light	**Moderate**	**Severe**	Dust haze	Visibility less than 4 km.	Visibility less than 2 km.	Visibility less than 1 km.
Frost	Few plants affected	Plants partially killed	Plants mostly killed	High winds	High: 30-50 km/hr	Very high: over 50 km/hr	

Code for weather at the time of observation	0- Fair (Clear & slightly clouded) 1-Vraiable sky	2- Mainly overcast 3- Sand storm or dust storm driftings now	4- Fog or thick dust haze 5- Drizzle	6- Rain 7- Snow or sleet	8- Shower (s) 9- Thunderstorms with or without precipitation

Period No.	Week No.	Date	Rain	Snow	Frost	Dew	Fog or Mist	Thunderstorm without hail	Hailstorm-mention diameter of hailstorm in cm.	Dust haze	Dust storm or squall	High winds	Remarks indicating times of occurrence and approximate duration of weather phenomena and their effects, if any on standing crops
			Brief description of weather during past 24 hours ending at 0700 hours (Local mean time)										
(1)	(2)	(3)	(4)	(5)	(6)	(7)	(8)	(9)	(10)	(11)	(12)	(13)	(14)

Signature of observer

Annexure-XV : Crop Weather Calender

State : Madhya Pradesh **Crop :** Wheat **Variety :** WH-14, HD 1553, Sujata, Lok-1, N-4 **Soil** : Red Alluvial, Black Cotton

Districts : Jhabua, Indore, Ujjain, Dhar **Irrigated** **Duration**: 120-150 days

		Weeks 44–51 (November–December)	Weeks 52–5 (December–January)	Weeks 6–10 (February–March)
Weather warning	Rain	>50 mm per day	>100 mm per day	>100 mm per day
	Wet spell duration	>50 mm for 3-4 day	>75 mm for 3-4 day	>50 mm for 3-4 day
	Weather condition	Cloudy weather	Cloudy weather	Cloudy weather
	Drought	20 days	20 days	20 days
	High winds	>50 km per hr.	>50 km per hr.	>50 km per hr.
	Temperature	Max. Temp. >30°C	Max. Temp. >30°C	Max. Temp. >30°C
	Hail storm		Hail storm	Hail storm
Weather conditions favourable for pest & diseases	Pests	Stem borer, aphids, termites, army worms, cut worms	Stem borer, aphids, termites, army worms, cut worms	Stem borer, aphids, termites, army worms, cut worms
	Weather	Cloudy weather	Cloudy weather	Cloudy weather
	Diseases	Rust	Rust	Rust, leaf spots, leaf blight
	weather	High humidity, cloudiness	High humidity, cloudiness	High humidity, cloudiness

		Oct	44	45	46	47	48	49	50	51	52	1	2	3	4	5	6	7	8	9	10		12
Normal phase-wise water requirement (mm)			62					155							80		127				46		
Weekly normal weather	Total rainfall (mm)		3.2	4.6	1.8	4.4	7.6	0.8	0.3	1.5	1.3	0.8	2.7	0.5	0.9	0.8	0.7	0.3	0.6	0.9	0.4	0.4	0.1
	Max. temp. °C		32.1	31.4	30.9	29.8	28.6	28.2	27.8	28.0	27.1	26.5	26.8	26.5	27.4	27.8	28.4	29.7	30.4	31.9	33.1	34.7	35.5
	Min. temp. °C		15.4	14.4	14.5	13.4	12.4	11.6	10.5	10.6	9.8	9.6	10.0	9.7	10.3	10.0	10.4	11.6	12.4	13.9	14.9	16.1	17.4
	Sunshine hrs.		9.5	9.1	9.1	8.8	8.6	8.6	8.6	8.9	8.7	8.8	8.8	9.2	9.3	9.5	9.6	10.3	10.4	9.4	9.2	9.3	9.5

Total=470

Life history and mean dates of important epochs of crop growth	Standard weeks
SOWING	44–47 (November)
GERMINATION AND SEEDLING	44–48 (November)
VEGETATIVE GROWTH	48–3 (November–January)
FLOWERING	3–5 (January)
GRAIN FORMATION	5–9 (January–February)
MATURITY	9–12 (February–March)

Standard weeks	44	45	46	47	48	49	50	51	52	1	2	3	4	5	6	7	8	9	10		12
Months	November				December					January					February				March		

Oct: Months column preceding November.

DDGM (Agrimet), IMD, Pune 2004.

Annexure-XVI : Standard metrological weeks and dates

Period	Week No.	Dates	Period	Week No.	Dates
1	1 Jan.	1 - 7	7	27 July	
	2	8 - 14		28	
	3	15-21		29	
	4	22-28		30	
	5 Jan. 29-4 Feb.		31	30-5 Aug.	
2	6 Feb.	5-11		32 Aug.	
	7	12-18		33	
	8	19-25		34	
	9 Feb. 26-4 March (A)*		35	27-2 Sept.	
	3 10 March	5-11		36 Sept.	
	11	12-18		37	
	12	19-25		38	
	13 26 March 1 April		39	24-30 Sept.	
	4 14 April	2-8		40 Oct.	
	15	9-15		41	
	16	16-22		42	
	17	23-29		43	
	18 30 April - 6 May		44	29-4 Nov.	
	5 19 May	7-13		45 Nov.	
	20	14-20		46	
	21	21-27		47	
	22 28 May - 3 June		48	26-2 Dec.	
	6 23 June	4-10		49 Dec.	3.9
	24	11-17		50	10-13
	25	18-24		51	17-23
	26 25 June -1 July		52	24-31 (B)*	

Annexure: XVII- Procedure for Supply / Obtaining Meteorological Data

Data request may kindly be made in the format and be sent to IMD office, Pune by post or email (ndc@imdpune.gov.in) with full information such as the *Type of data* (Daily, weekly or monthly Rainfall, Daily or Monthly Surface data or Upper-Air data, etc,.), *Period* (From and To year or month etc.,), the *Region/Stations* (State-wise, District-wise or Station-wise) for which the data requested and the *Medium* (Floppy/CD/Hardcopy etc.,) on which the data is to be supplied. Additional information on the type of the project or research study with some brief background of the Institution/Company may also be given.

Based on the request received, National Data Cenetre, IMD, Pune would intimate the cost, availability of the data requested alongwith a format of *Undertaking Certificate* on the Data Utilisation. The cost for the data supply is accepted only as Demand Draft on any Nationalised Bank, favouring "Additional Director General of Meteorology (Research)", payable at Pune.

On receipt of all the required formalities from the party, the requested data are supplied alongwith the receipt for the DD amount. The cost estimation of data is based on the Standard Tariff approved by the Department. However, the cost of the data would be intimated by the Department to the part based on the availability of data.

All Data requests are to be addressed and sent to: Additional Director General of Meteorology (Research), India Meteorological Department, Shivajinagar,PUNE 411 005

Whom to contact

Director-in-charge, National Data Centre

India Meteorological Department, Shivajinagar, Pune - 411 005

Source: **IMD, New Delhi**

Annexure-XVIII : Pan Coefficient (Kp) of Class A Pan for different levels of mean relative humidity and wind speed under cropped and fallow area

Class A PAN	Case 1 : Pan placed in short green cropped area				Case 2 : Pan placed in dry fallow area			
Mean RH (%)		Low < 40	Medium 40-70	High < 70		Low < 40	Medium 40-70	High > 70
Wind speed (Km/day)	Wind ward distance from green crop (m)				Wind ward distance from dry fallow (m)			
Light < 175	1	0.55	0.65	0.75	1	0.70	0.80	0.85
	10	0.65	0.75	0.85	10	0.60	0.70	0.80
	100	0.70	0.80	0.85	100	0.55	0.65	0.75
	1000	0.75	0.85	0.85	1000	0.50	0.60	0.70
Moderate 175-425	1	0.50	0.60	0.65	1	0.65	0.75	0.80
	10	0.60	0.70	0.75	10	0.55	0.65	0.70
	100	0.65	0.75	0.80	100	0.50	0.60	0.65
	1000	0.70	0.80	0.80	1000	0.45	0.55	0.60
Strong 425-700	1	0.45	0.50	0.60	1	0.60	0.65	0.70
	10	0.55	0.60	0.65	10	0.50	0.55	0.65
	100	0.60	0.65	0.70	100	0.45	0.50	0.60
	1000	0.65	0.70	0.75	1000	0.40	0.45	0.55
Very strong >700	1	0.40	0.45	0.50	1	0.50	0.60	0.65
	10	0.45	0.55	0.60	10	0.45	0.50	0.55
	100	0.50	0.60	0.65	100	0.40	0.45	0.50
	1000	0.55	0.60	0.65	1000	0.35	0.40	0.45

❑❑❑

FSC
www.fsc.org
MIX
Papier | Fördert
gute Waldnutzung
FSC® C083411

Zeitfracht Medien GmbH
Ferdinand-Jühlke-Straße 7
99095 Erfurt, Deutschland
produktsicherheit@kolibri360.de